IMAGES
of America

CONOWINGO DAM

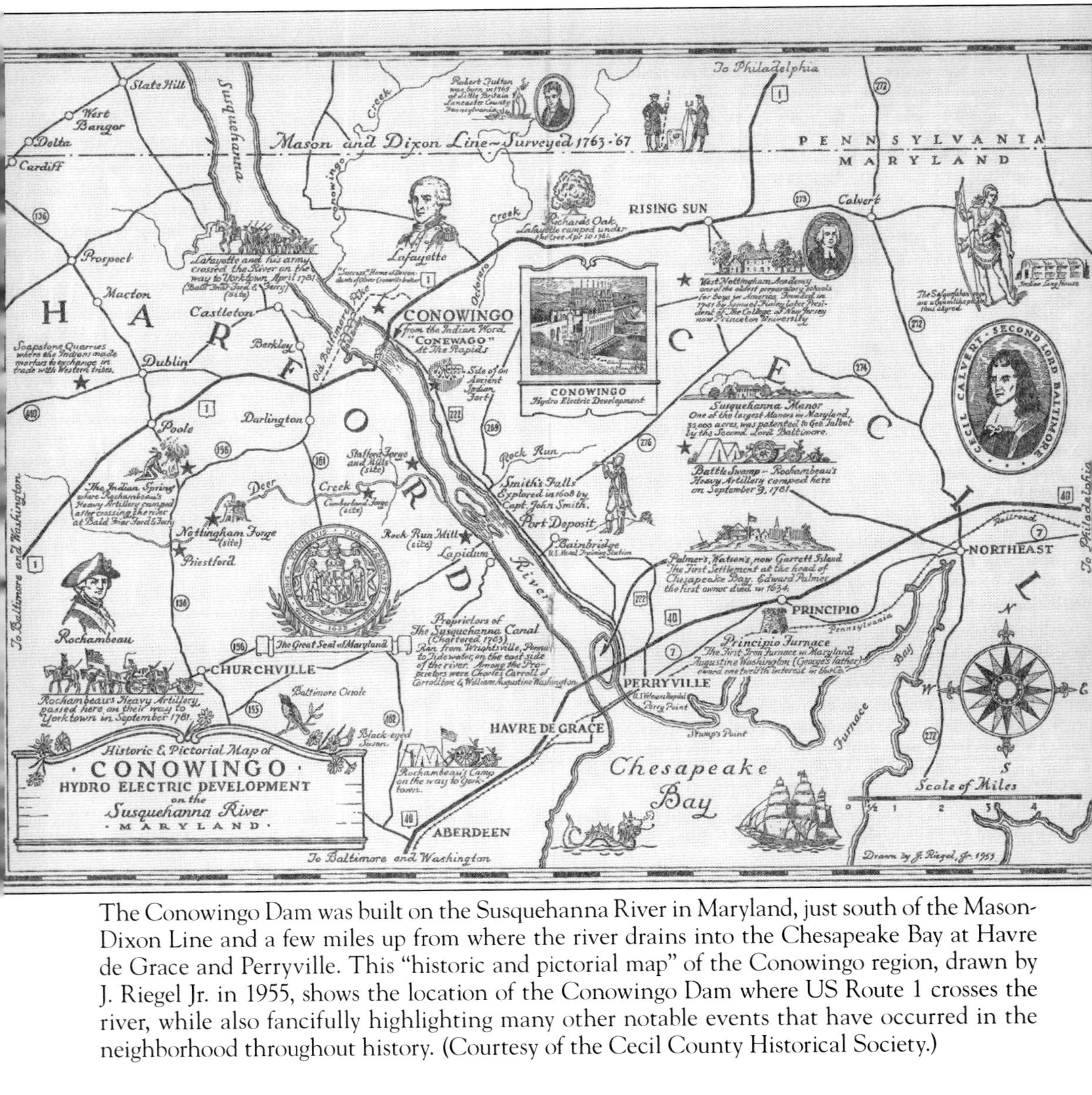

The Conowingo Dam was built on the Susquehanna River in Maryland, just south of the Mason-Dixon Line and a few miles up from where the river drains into the Chesapeake Bay at Havre de Grace and Perryville. This "historic and pictorial map" of the Conowingo region, drawn by J. Riegel Jr. in 1955, shows the location of the Conowingo Dam where US Route 1 crosses the river, while also fancifully highlighting many other notable events that have occurred in the neighborhood throughout history. (Courtesy of the Cecil County Historical Society.)

ON THE COVER: At the time of its completion, the Conowingo hydroelectric plant was the second-largest in output capacity ever built in the United States. The erection of it was an immense undertaking that required the impounding of a mighty river and the efforts of thousands of construction laborers, technicians, and hydro and civil engineers. This 1929 photograph shows the nearly mile-long dam shortly after its generators began to produce power. (Courtesy of the Conowingo Dam.)

IMAGES
of America

CONOWINGO DAM

John R. Paulson with Erin E. Paulson

ARCADIA
PUBLISHING

ISBN 978-1-4671-2617-5

Published by Arcadia Publishing
Charleston, South Carolina

Printed in the United States of America

Library of Congress Control Number: 2017931405

For all general information, please contact Arcadia Publishing:
Telephone 843-853-2070
Fax 843-853-0044
E-mail sales@arcadiapublishing.com
For customer service and orders:
Toll-Free 1-888-313-2665

Visit us on the Internet at www.arcadiapublishing.com

Contents

Acknowledgments

I am grateful to Erin Paulson, my writing collaborator, for her valuable contributions to the text. Her observant style and warm tone simultaneously sharpened and leavened many of its passages.

Mike Dixon, historian from Cecil County, Maryland, opened my eyes to the cultural history of the 1920s and the Conowingo Dam's impact on the region, which broadened the scope and enriched the substance of this book. He also proofread the text with an eye for its historical accuracy.

At the Conowingo Dam, Don Baldwin and Fred Smith made understandable to me the mystifying jargon of the civil, electrical, and hydro engineering fields, the specialized processes of building the dam, and the complexities of how it works. They also made available the extensive archive of photographs housed at the dam to make this book possible. Unless otherwise noted, all images in this book are from the photographic collection of the Conowingo Dam.

Maryland Public Television commissioned the documentary from which this book evolved, as such I am indebted to Steven Schupak, whose commitment to documenting Maryland history and culture is impressive; Michael English, whose sound judgment gave the documentary structure and shape; and Patrick Ridgely, whose keen eye and research acumen resulted in many of the photographs seen herein.

Longtime friend James Arntz provided sensitive suggestions for clearing the sludge from the manuscript.

Lastly, I wish to express gratitude to Dawn, Sean, and Erin for their ceaseless love and support, and for giving me permission to try on a new hat.

INTRODUCTION

Conowingo Dam was a place I had never heard of. I only learned about it when Maryland Public Television asked me to produce an hour-long documentary, *Conowingo Dam: Power on the Susquehanna*. As a filmmaker, I am by nature a generalist, and when a new project like Conowingo comes along, I become immersed in the film's subject matter—doing research, conducting interviews, and collecting everything I need to tell the story.

The first thing I discovered is that there are few public histories of the Conowingo Dam. I was doing primary research for a story that had not yet been told. Our production team scoured regional newspaper archives, surveyed the collections of historical societies, and browsed the holdings of the Exelon Corporation, the power company that now owns the Conowingo Dam. From out of deep storage, we unearthed a trove of photographs few people had ever seen, dating to the dam's inception in the early 1920s.

The birth of the Conowingo Dam had been extensively photographed, with every step in the construction process preserved for posterity, both at ground level and in stunning vistas from the sky. This rich collection of visuals enabled our television program not only to track the construction of the dam but also to paint a picture of the political battles fought over it, and to document the impact of the project on what had been a sparsely populated region of Maryland.

Looking now at the handsome dam as it crosses the Susquehanna River, it is difficult to visualize the momentous effort that went into the planning and construction or the remarkable machinery housed within that cement facade. I was drawn particularly into the fine details of photographs showing the intricate, and oftentimes mystifying, construction processes and equipment. I also found many of the images, especially the panoramic photographs taken by R. Clements, to be aesthetically beautiful. And I was completely engaged by the old newspaper articles that vividly recount the dam's milestones, successes, and setbacks.

The sheer manpower needed to build Conowingo Dam is astounding and unlike anything you would see today. The photographs capture a small army of civil, electrical, mechanical, and hydro engineers; surveyors and crane operators; dynamite blasters and concrete puddlers; carpenters; ironworkers; welders; plumbers; pipefitters; electricians; and manual laborers whose efforts altered the path of the Susquehanna River so that America could have more electricity. They must have known they were building something important and that it was going to do a great amount of good for the nation in the years to come.

The colossal construction scene was alive with the din of heavy machinery sounding throughout the river valley: steam shovels puffing smoke from smokestacks, jackhammers gouging out the bedrock, concrete mixers churning cement, cranes hoisting heavy materials high overhead, and locomotives chugging along the temporary rail lines to carry in supplies. For two years, a swarming mass of men and machines were in perpetual motion. Their mission of impounding a major waterway and controlling so much raw power represents the very meaning of the American can-do spirit.

Looking back from today's more advanced technological era, I was particularly intrigued by the construction processes that were so innovative for their time. The building of hydropower facilities, like all engineering disciplines, requires very specialized knowledge and skill. What remarkable combination of engineering, hydrography, design, and construction know-how was employed to erect a hydroelectric dam in 1926–1928, before heavy mechanization and computer automation? I hoped to demystify the technology of hydro dam construction (the methods and machinery used to build it), to explain the unique components (generators, turbines, electrical controls, and power transformers) that make it run, and to provide a guided tour as to how Conowingo Dam turned flowing water into electricity.

I realized rather quickly that the material I had found for the television show could also make an interesting book. Stories are reduced to their essence in television, and the images are fleeting. With a book, you can set your own pace, and if you wish to pull out your magnifying glass and peer into the crannies of the half-built powerhouse, or dawdle over the photograph of a boy atop an iceberg in downtown Port Deposit, Maryland, or delight in the view of the drained Susquehanna River, its rocky bottom laid bare by the cofferdams (a sight that may never be seen again), you may do so.

Capt. John Smith was the first European to explore the Susquehanna, in 1608, and he praised the beauty of the setting where the hydroelectric project was to emerge centuries later: "Heaven and earth never seemed to have agreed better to frame a place for Man's commodious and delightful living." Smith met the Susquehannochs on his two voyages up the Chesapeake Bay and discovered that the Susquehanna River Valley was already a natural artery for travel among the Native American tribes.

As the Maryland colony became firmly established, the river continued as a great crossroads of people and commerce. During the Revolutionary War, the Marquis de Lafayette crossed the river at Bald Friar Ford on his way south to play a decisive role in the victorious Battle of Yorktown. One of his officers referred to the ford as "diabolique" because of the slippery rocks and potholes that injured his horses and wrecked his wagons. In the 1800s, the Susquehanna Canal ferried goods from the head of the Chesapeake Bay into Pennsylvania and back. The first Congress of the United States even considered the river an appropriate site for the federal capital, debating whether the Potomac River, the Delaware River, or the Susquehanna River afforded a better connection to the Northwest.

The dam ultimately built at Conowingo, Maryland, came during a dynamic time in American history. World War I had just ended, and the availability of electric power was transforming the country. The Conowingo Dam was part of the race to harness the nation's waterways to meet the new demands for power, and soon after its completion in 1928, the federal government began actively promoting and funding new and even greater hydropower projects. As Pres. Franklin Roosevelt pronounced in 1934, standing before the newly erected dam at Fort Peck, Missouri, "We are going to make every ounce and every gallon of water that falls from the Heaven and the hills count."

In making the television documentary, and now this book, I was particularly struck by what occurred in that rural river valley so that the Conowingo Dam could be built: the men who labored to build it, the ones who sacrificed their lives for it, the massive components they installed inside the dam, the little village of Conowingo that vanished because of it, and the raging floodwaters that tested it. This book explores the remarkable story of laborers, managers, planners, engineers, politicians, and financiers who are now long gone, but whose vision and dedication can be seen almost 90 years later in the powerful ribbon of concrete that stretches so gracefully across the Susquehanna River.

One

Why Conowingo?

The Conowingo Dam was built during the Roaring Twenties, when the inventions of Thomas Edison and the power-distribution innovations of George Westinghouse were changing American life. Homes were illuminated and becoming equipped with newfangled electrical conveniences like the refrigerator, washing machine, telephone, and radio. In cities, streetcars were electrified, and the country's populace was increasingly on the move. The use of electricity was doubling every six years.

This rapid growth sent engineers in search of more and cheaper supplies of power, and for that, they turned to a valuable natural resource, the force of flowing water. The Federal Water Power Act, passed in 1920, gave the national government a role in coordinating the inevitable wave of hydroelectric projects that would evolve. Dams like the Conowingo generated electric power by impounding rivers and funneling water currents past massive turbine blades, causing them to spin. This kinetic energy was transformed by immense generators into electrical energy, which was then transmitted to nearby factories, farms, cities, and towns. But these hydroelectric dams could not be built just anywhere; there were certain essential prerequisites for the appropriate location.

A potential building site, for example, required a drop in elevation sufficient to produce the water flow necessary to drive the giant waterwheels, as well as topographical and geologic features suitable for supporting the massive structure. And apart from the requirements imposed by geography, there also had to be a nearby market of customers who would purchase the generated energy, not to mention optimistic investors who could justify the enormous construction costs and believe in its ability to bring them a positive rate of return. Finally, as with all such major endeavors, the politicians and regulators had to be won over.

In 1925, these many stars aligned for constructing a dam on the lower Susquehanna River near the quiet village of Conowingo, Maryland. And in March 1926, work began on the mammoth $52 million structure, which would impound the ancient river and, when completed, be second in size only to the enormous hydroelectric facility at Niagara Falls, New York.

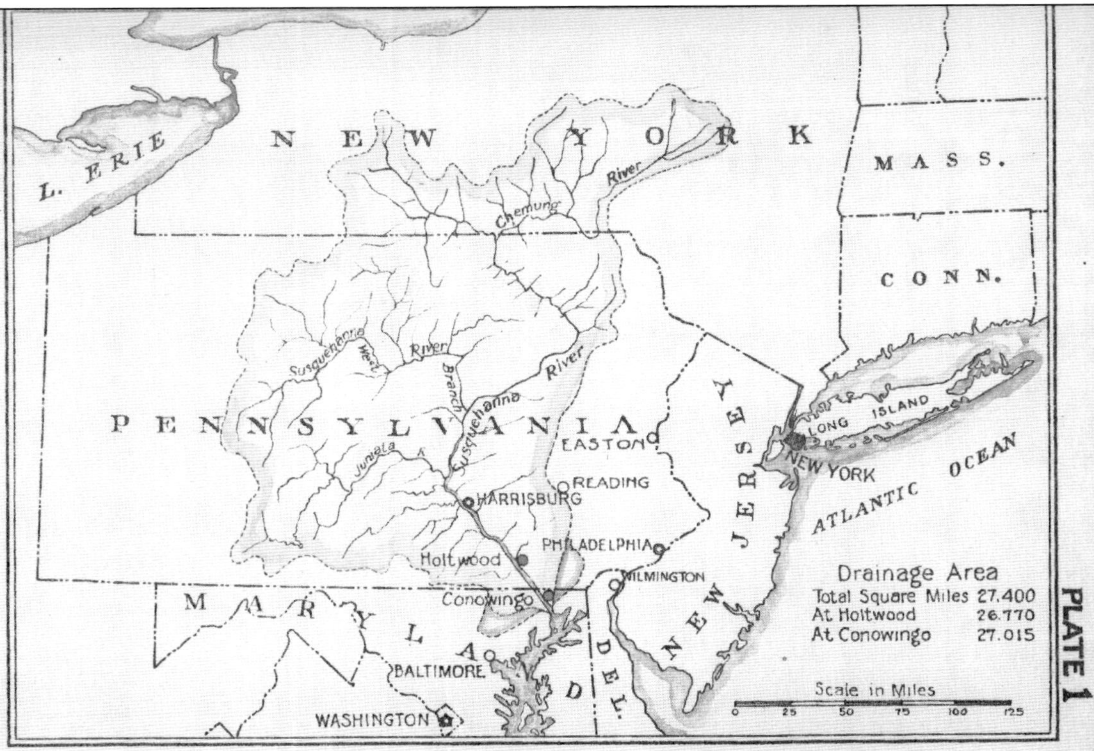

SUSQUEHANNA RIVER DRAINAGE AREA

The Susquehanna is the longest river in the northeastern United States and one of the oldest rivers in the world. The figure above shows an immense watershed with a drainage area of 27,400 square miles. The river rises in New York State, flows down through Pennsylvania, and finishes in Maryland, where it empties into the Chesapeake Bay. For its first 400 miles, the Susquehanna River flows southward on a gentle slope. But when it leaves the fertile farmlands of southeastern Pennsylvania and crosses the Mason-Dixon Line into Maryland, the current accelerates as it drops a dramatic 208 feet in 27 miles, making the last stretch ideal for the generation of electricity. Interest in building dams along the lower Susquehanna River developed in the late 1890s, when a group of Marylanders began the acquisition of lands along the river; and by 1920, several small-scale stations, such as the Holtwood Dam in Pennsylvania, generated power there. But until the mid-1920s, there was neither adequate financing nor a market for the kind of large-scale plant that was feasible.

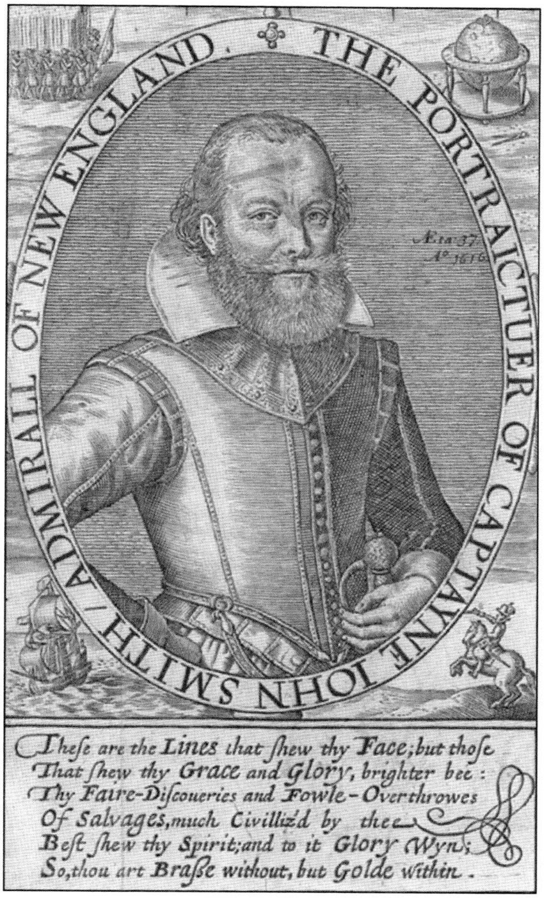

The lower Susquehanna River is rich in history, having always been an active area of human habitation. When English explorer Capt. John Smith sailed up the Chesapeake Bay in 1608, he reached the head of the bay and continued up the Susquehanna River, where he discovered the rich natural wonders of the region as well as abundant native peoples living along its shores and fishing in its waters. He wrote, "Upon this river inhabit a people called Susquehannock," who greeted him "with skins, bows, arrows, targets, beads, swords, and tobacco pipes for presents. Such great and well-proportioned men are seldom seen, for they seemed like giants to the English." As Smith steered farther up the Susquehanna River, his boat was stopped by the very shallow rapids near present-day Conowingo, and today, these rapids are still known as Smith's Falls. (Both, courtesy of the Library of Congress.)

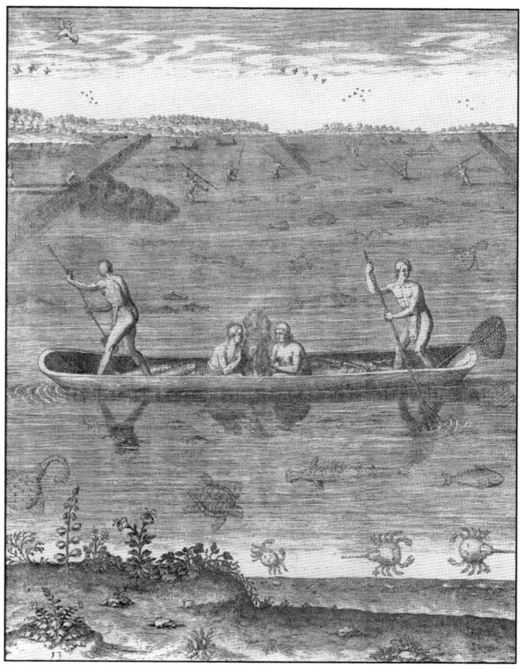

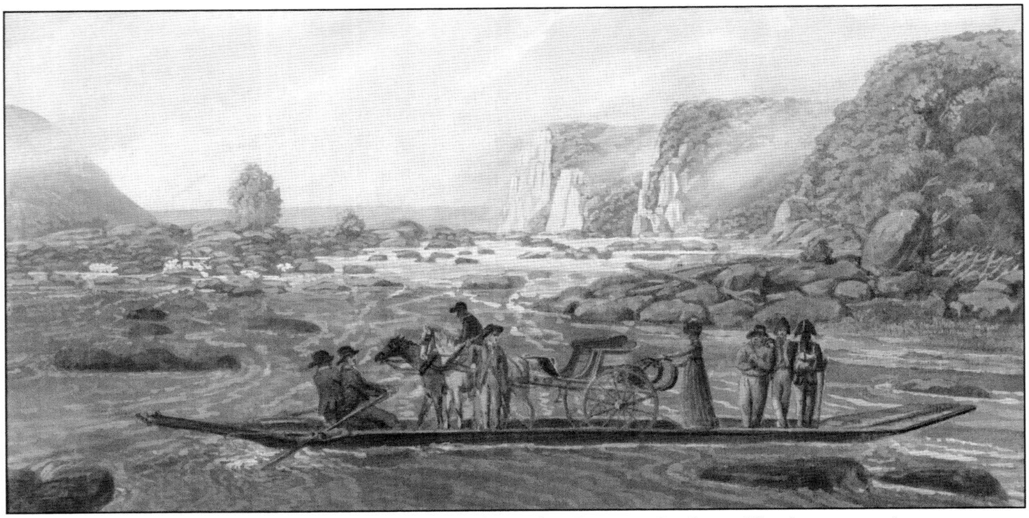

In colonial times, travelers on roads that ran along the Eastern Seaboard had to cross the narrowest and swiftest section of the lower Susquehanna aboard ferryboats, such as seen here. During the American Revolution, General Lafayette took his French army south across the Susquehanna River at Bald Friar Ford, near the eventual damsite, on his way to Yorktown. In the mid-1800s, as escaping slaves crossed from Maryland to free soil in Pennsylvania, the Underground Railroad had many clandestine crossing points in this part of the river. (Courtesy of Internet Archive.)

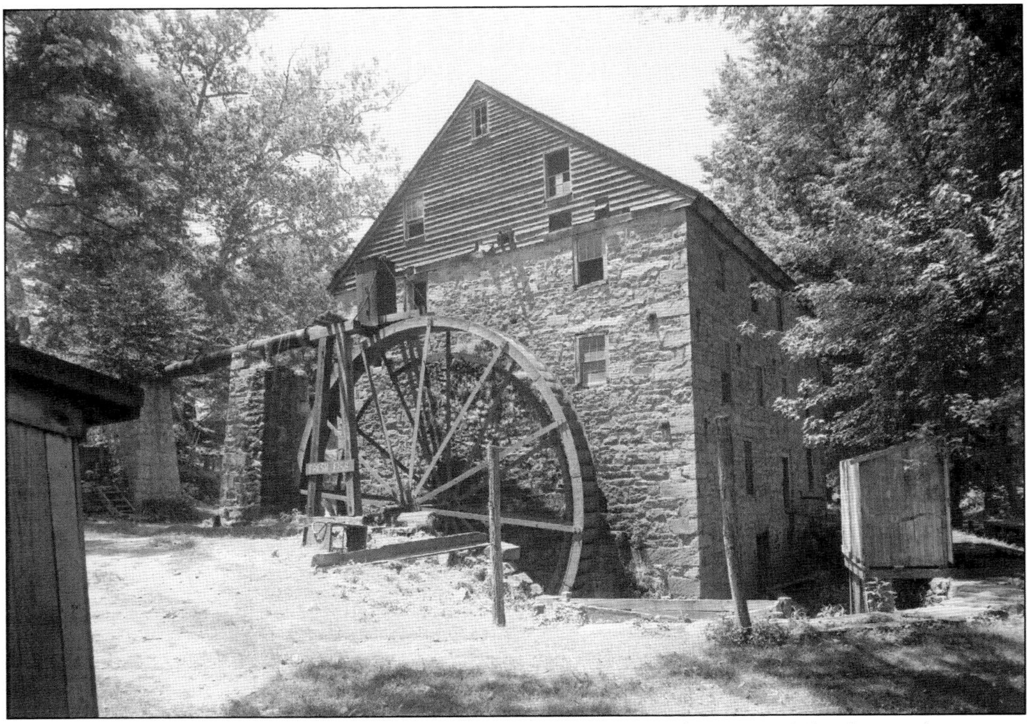

Colonial Americans turned to the lower Susquehanna River for power. The force of its water turned the wheels of gristmills that dotted the riverbanks, such as the Rock Run Gristmill, seen here, erected by John Stump in 1798. During its prosperous years in the early 19th century, Rock Run Gristmill ground flour for both local and international markets. (Courtesy of the Library of Congress.)

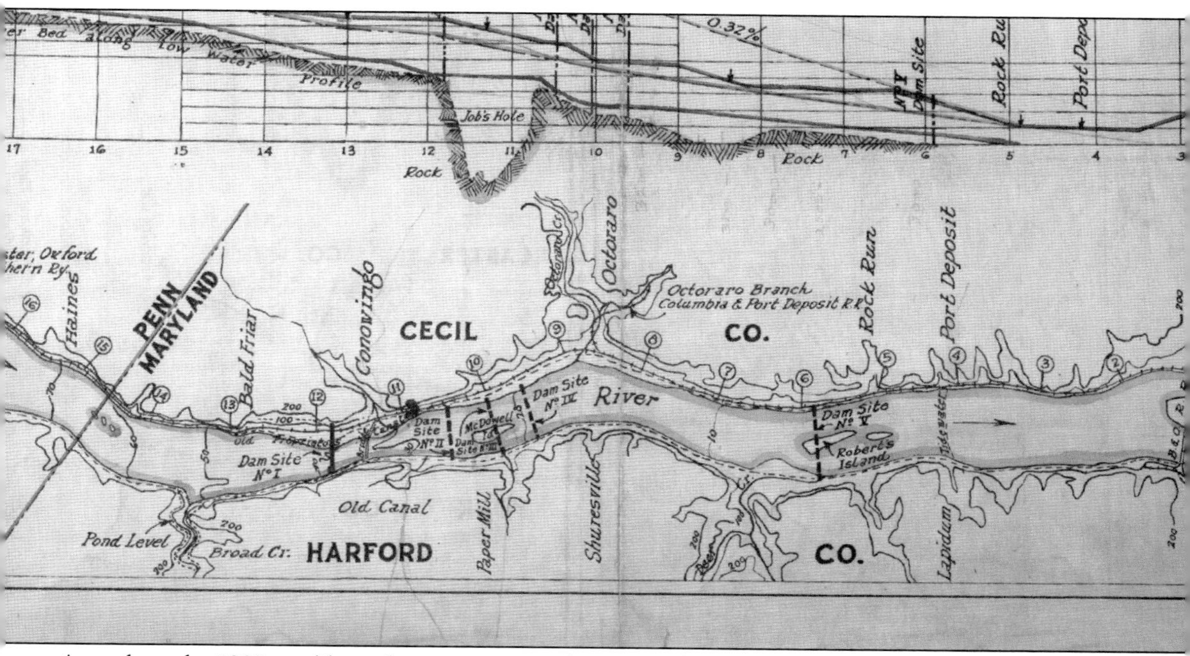

As early as the 1880s, public utility managers and investors from both Philadelphia and Baltimore looked to the lower Susquehanna River to meet the growing demand for electricity. Speculators bought riverside land near Conowingo, while surveyors and hydro engineers searched for the spot possessing the greatest promise for power production. These graphs were taken from the 1923 feasibility study, "Report on Conowingo Hydro-Electric Development," in which five possible sites for a dam were studied. For each location, engineers took into account topography, historical data from the US Geological Surveys, and hydrographs of the river's streamflow with the goal of locating the dam where the transformation of water current into electric current would be maximized. Economists forecast the costs of constructing, operating, and maintaining the facility and predicted how much electricity would be available at each site, with the goal of convincing politicians, public utility managers, and financiers of the fiscal feasibility of the project.

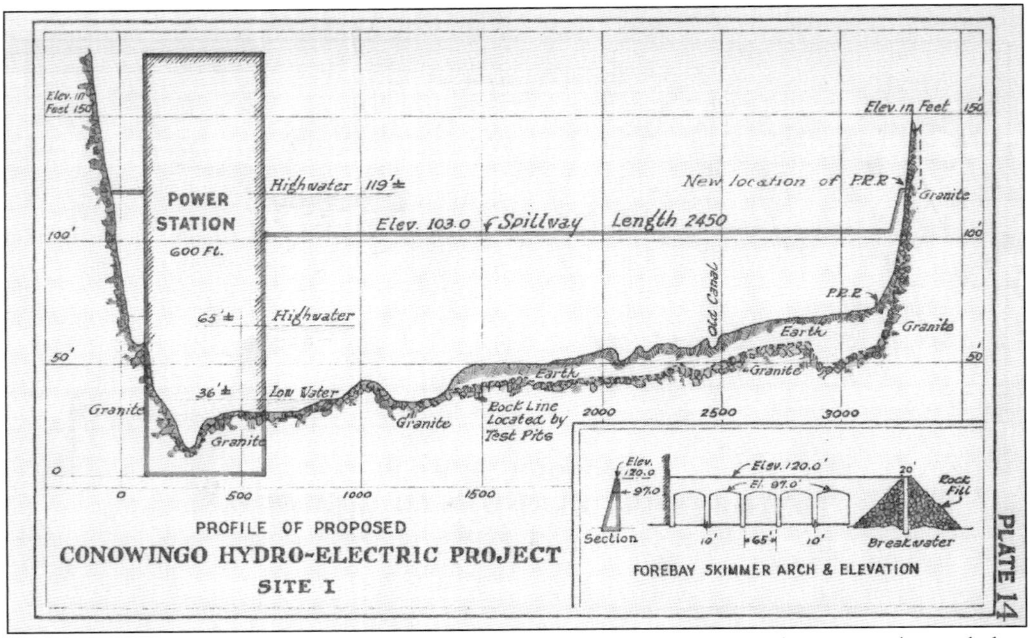

This plate shows a cross section of the Susquehanna River at Site I with a main channel that ran along the river's west bank. This is where the current is strongest and where the river's fall would generate the greatest power. The report proposed a 600-foot-long powerhouse to hug the western end of the facility and a 2,450-foot-long spillway to run eastward from the powerhouse to the far bank of the river.

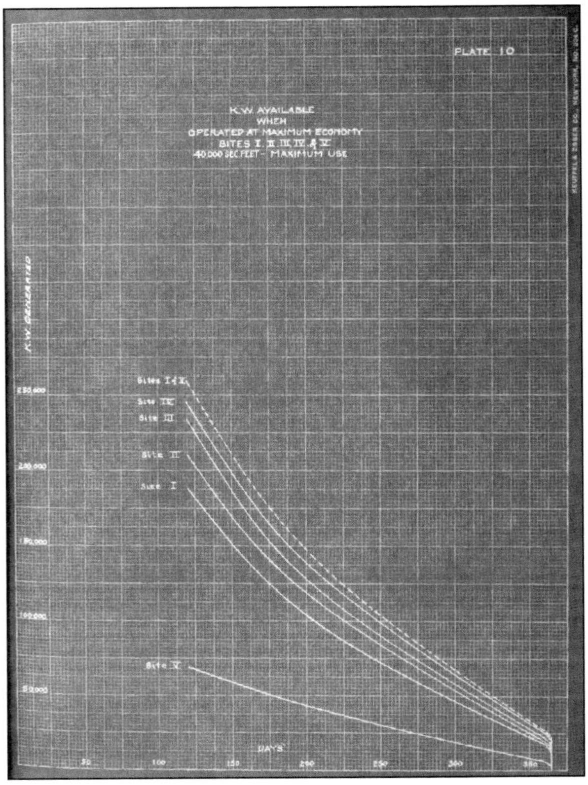

Hydro engineers speak in terms of *effective head* to describe a site's prospects for power generation. It is a measure of the river's fall from an elevation above a dam to an elevation below it, and therefore quantifies the potential energy a hydroelectric dam can harness. This graph shows the effective head as calculated for the four proposed damsites, with Site IV providing the greatest.

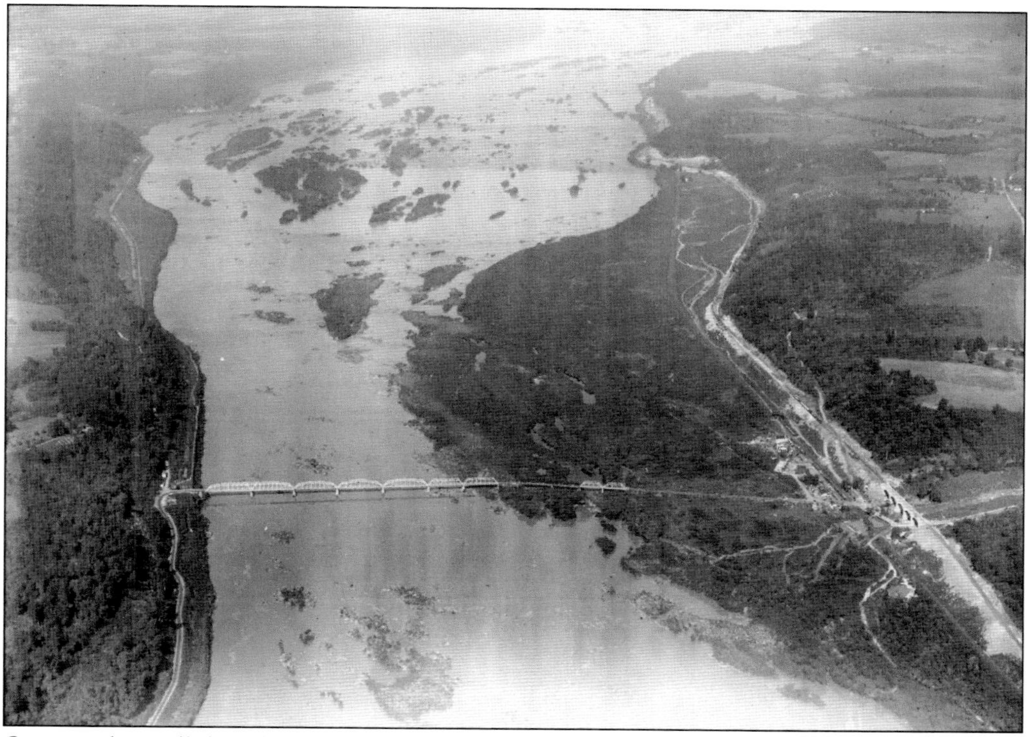

One site above all the others held the greatest promise of power. It was located nine miles upriver from where the Susquehanna River emptied into the Chesapeake Bay, at the little Maryland village of Conowingo (from a Susquehannock word meaning "at the rapids"). This aerial view from the 1920s shows Conowingo nestled into the east bank of the river near the very rapids that stopped Capt. John Smith in 1608. (Courtesy of the Harford County Historical Society.)

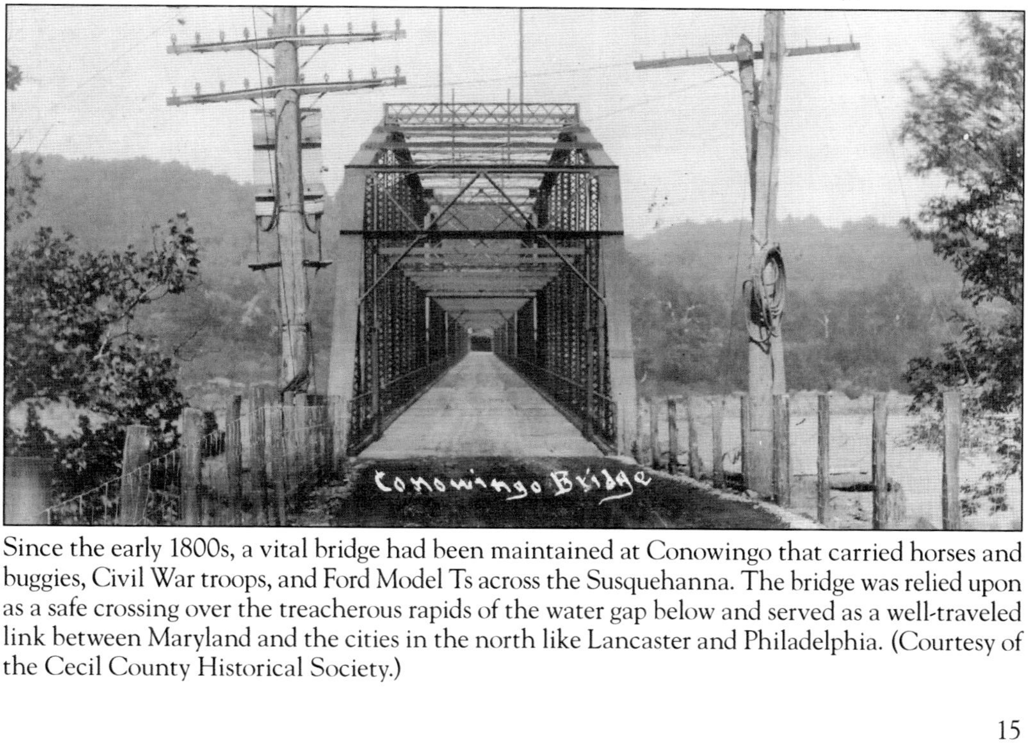

Since the early 1800s, a vital bridge had been maintained at Conowingo that carried horses and buggies, Civil War troops, and Ford Model Ts across the Susquehanna. The bridge was relied upon as a safe crossing over the treacherous rapids of the water gap below and served as a well-traveled link between Maryland and the cities in the north like Lancaster and Philadelphia. (Courtesy of the Cecil County Historical Society.)

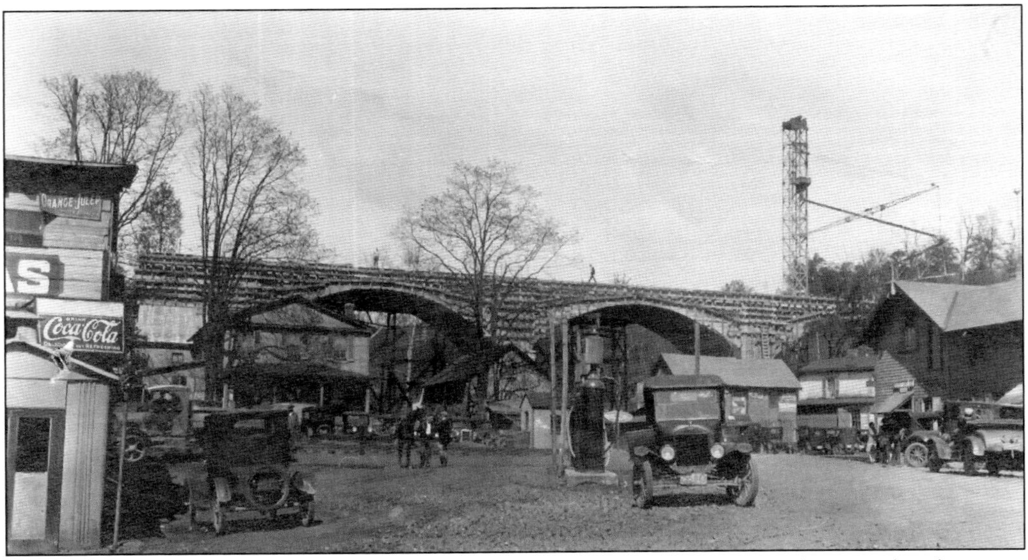

Conowingo was settled in the 1700s, and during the 1800s, it was a regular stop for changing horses on the stage run between Baltimore and Philadelphia. In the 1830s, the newly constructed railroads brought growth and industry to the town, and by the mid-1920s, Conowingo bustled with 350 residents, a blacksmith and harness maker, vegetable and seafood canneries, and several sawmills and churches. (Courtesy of the Cecil County Historical Society.)

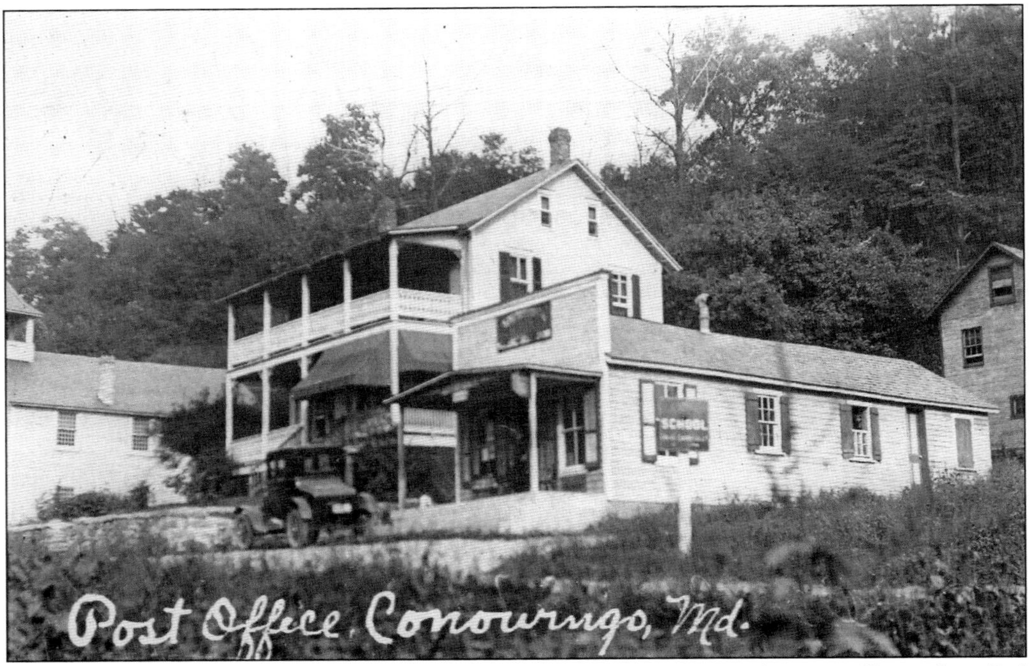

Doc Ragan's general store, shown here, was the heart of the town. As the post office, it handled the mail; as a drugstore, it sold pills and liniments; as the general store, it stocked groceries, fishing tackle, cooking utensils, and dry goods—a little bit of everything. But the proximity of the village to the river doomed all such establishments. When the power dam was completed and the floodgates slammed shut, the river pooled behind it, and the town of Conowingo was submerged. (Courtesy of the Cecil County Historical Society.)

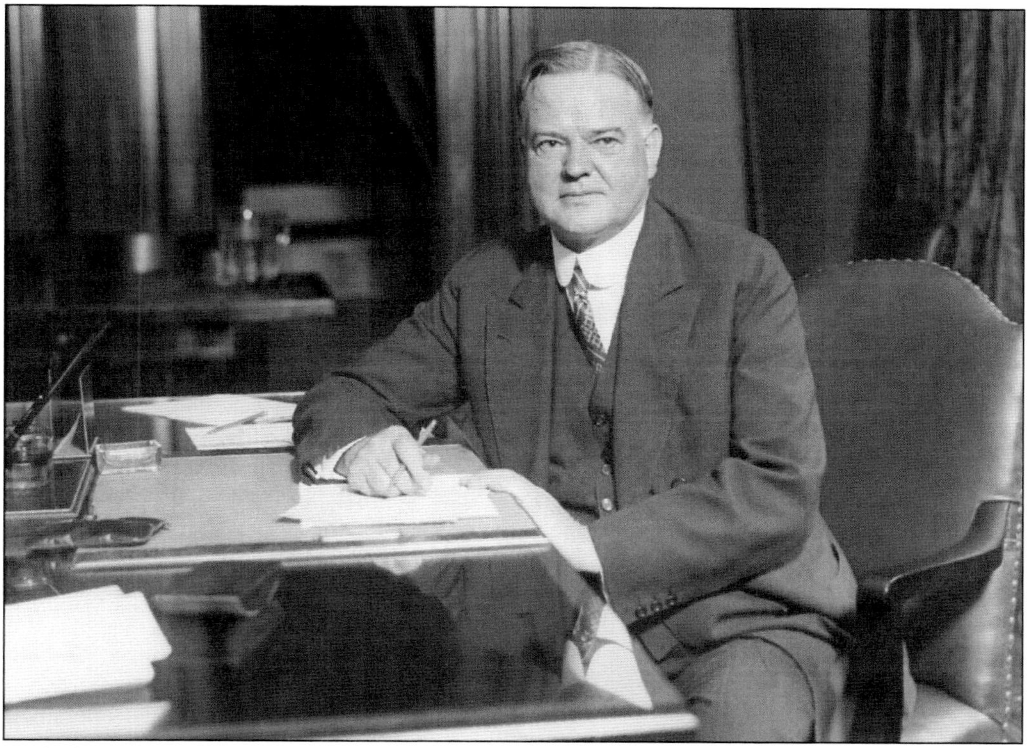

With the Conowingo site chosen, the approval of politicians and regulators was essential, and US secretary of commerce Herbert Hoover declared his support as follows: "This is a particularly fortunate time for the launching of such a project. Money is obtained easily through the stock markets. This must be developed without delay." A key factor in influencing Hoover's enthusiasm was that Conowingo Dam was to be entirely privately financed without federal investment. (Courtesy of the Library of Congress.)

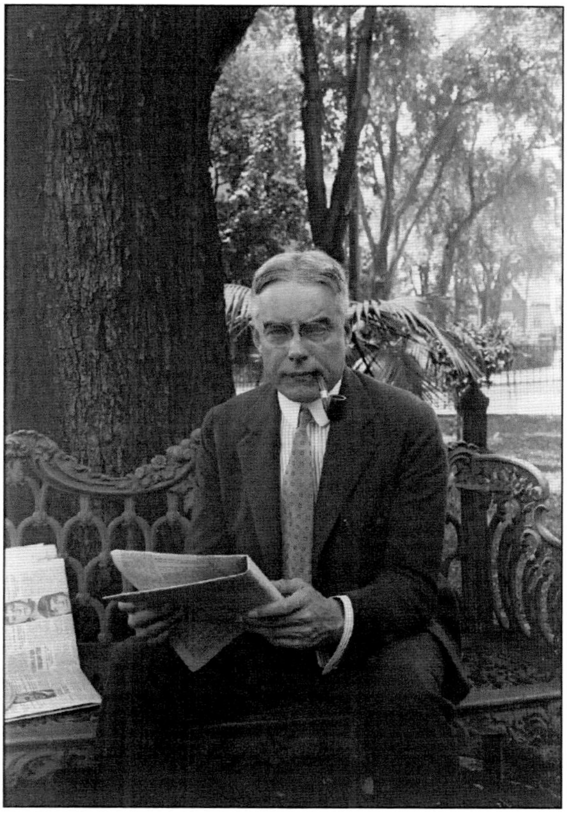

The governor of Maryland, Albert Ritchie, declared, "The authorization . . . of the Conowingo project has my complete approval. This is the only possible way in which this great natural resource can be developed. The construction work will give employment to thousands of Maryland people, will add greatly to the taxable bases of Harford and Cecil Counties and will measurably increase the revenues of the state." (Courtesy of the Library of Congress.)

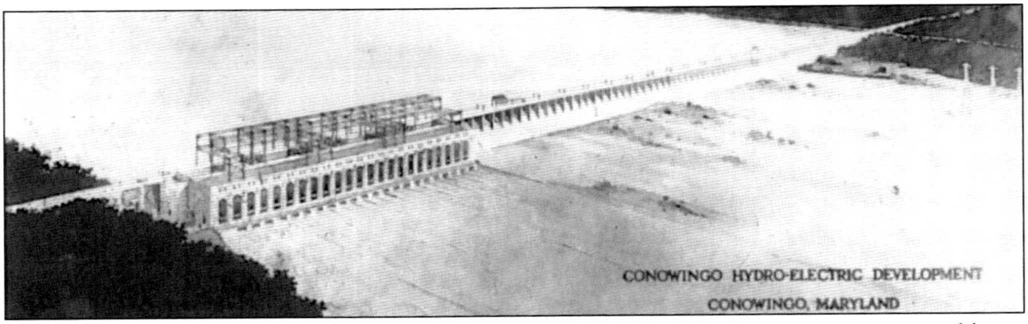

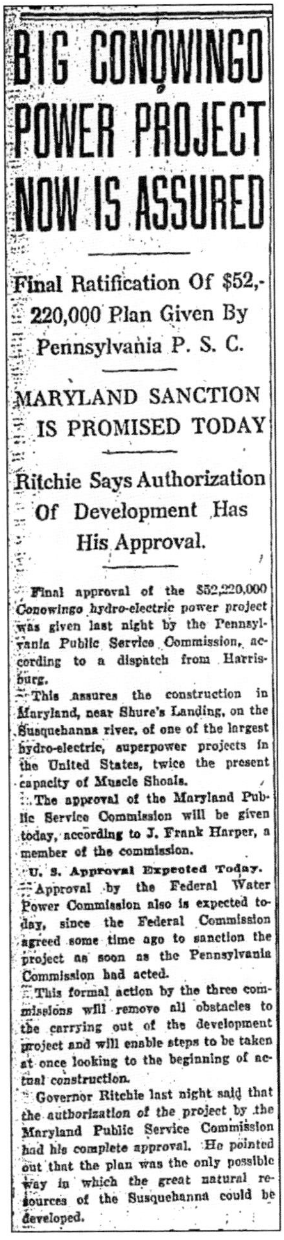

BIG CONOWINGO POWER PROJECT NOW IS ASSURED

Final Ratification Of $52,220,000 Plan Given By Pennsylvania P. S. C.

MARYLAND SANCTION IS PROMISED TODAY

Ritchie Says Authorization Of Development Has His Approval.

Final approval of the $52,220,000 Conowingo hydro-electric power project was given last night by the Pennsylvania Public Service Commission, according to a dispatch from Harrisburg.

This assures the construction in Maryland, near Shure's Landing, on the Susquehanna river, of one of the largest hydro-electric, superpower projects in the United States, twice the present capacity of Muscle Shoals.

The approval of the Maryland Public Service Commission will be given today, according to J. Frank Harper, a member of the commission.

U. S. Approval Expected Today.

Approval by the Federal Water Power Commission also is expected today, since the Federal Commission agreed some time ago to sanction the project as soon as the Pennsylvania Commission had acted.

This formal action by the three commissions will remove all obstacles to the carrying out of the development project and will enable steps to be taken at once looking to the beginning of actual construction.

Governor Ritchie last night said that the authorization of the project by the Maryland Public Service Commission had his complete approval. He pointed out that the plan was the only possible way in which the great natural resources of the Susquehanna could be developed.

By January 1926, the wheels of progress were turning inexorably. The Federal Water Power Commission had sanctioned the dam, and the Maryland Public Service Commission granted the license to build it. The commission's chairman, Ezra B. Whiteman, proclaimed, "This plant will be one of the greatest units in an Eastern superpower system," which included the dams at Niagara Falls and in Muscle Shoals, Alabama. The *Baltimore Sun* announced that Conowingo would be "one of the largest superpower projects in the United States," and went on to boast that "such a structure, it is declared, will make the Conowingo site a scenic beauty of the east." The big Conowingo power project was assured, and on March 8, 1926, work began on the privately funded $52,000,000 power plant across the Susquehanna River. (Left, courtesy of the *Baltimore Sun*.)

Two

Dam Construction Begins

The Conowingo Dam was built on a Maryland river, yet it was financed and owned by the Philadelphia Electric Company in Pennsylvania. Power generated in Maryland would be sent far to the north to operate Philadelphia's streetcars, which had rapidly changed over from horses to electric power. This arrangement was opposed by some of the residents of Maryland, who thought Conowingo power should be sent to Baltimore instead.

These feelings would grow into a central issue of the campaign for the Maryland governorship in the fall of 1926, but at the beginning of that year, there had been no obstacles to the Conowingo plan. The Philadelphia Electric Company awarded a contract to the Stone & Webster Company of Boston to design the dam and build the powerhouse, and the Arundel Corporation of Baltimore was hired to construct the dam's spillway and a new highway bridge on its deck. By the spring of 1926, work was under way on this ambitious plan to harness the Susquehanna River for the public good, and most particularly for the residents of Philadelphia.

The design called for a powerhouse with seven 54,000-horsepower generators, for a total initial capacity of 378,000 horsepower, the largest amount of power that had ever been installed in a single step. The design also provided for future installation of four additional generators, which when completed, would give the plant a total capacity of approximately 600,000 horsepower. In an average year, the Conowingo plant would deliver 1,250,000 kilowatt hours to the public.

As a concrete gravity dam, the Conowingo would sit under its own weight on Susquehanna River bedrock. A massive spillway section with 50 flood-control gates called crest gates would run from the powerhouse to the opposite shore with a 220-kilovolt transformer and transmission system on its roof. Roughly 1,000 steel towers and an extraordinary length of power lines would extend to Philadelphia, bringing sorely needed electricity to that market. When completed, the new dam would stretch nearly a mile shore-to-shore and create a 14-mile-long, 9,000-acre reservoir behind it called Conowingo Pond.

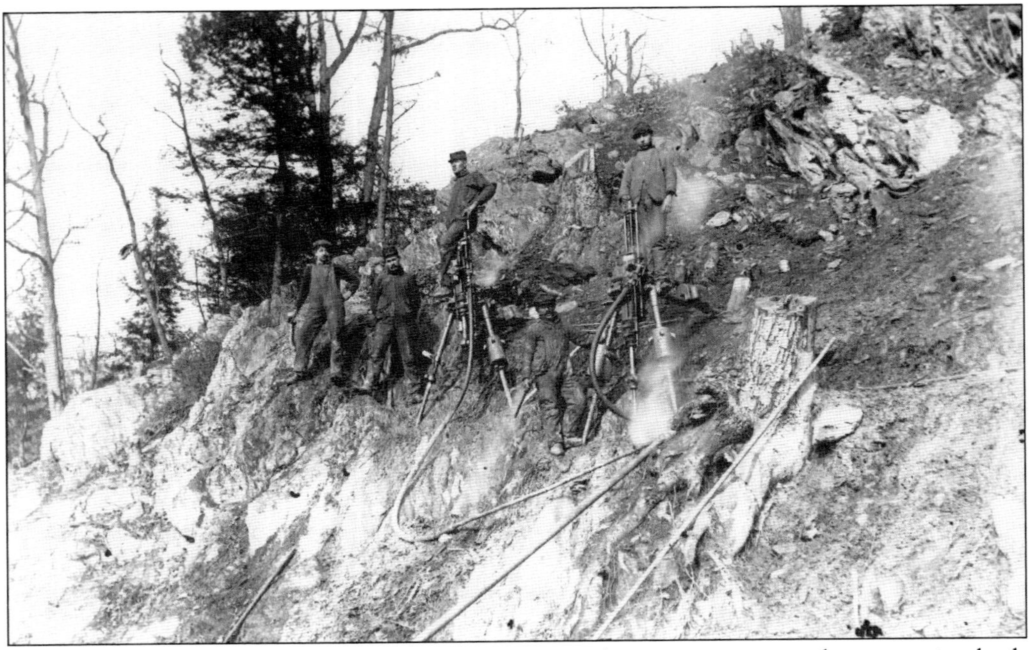

The initial step in the building process was to prepare the construction site by excavating both the granite riverbed on which the dam would sit and the steep rocky abutments where each end of the dam would be anchored. This image shows workers in the earliest days of the operation using pneumatic hammers to prepare the hillside for the abutments. (Courtesy of the Cecil County Historical Society.)

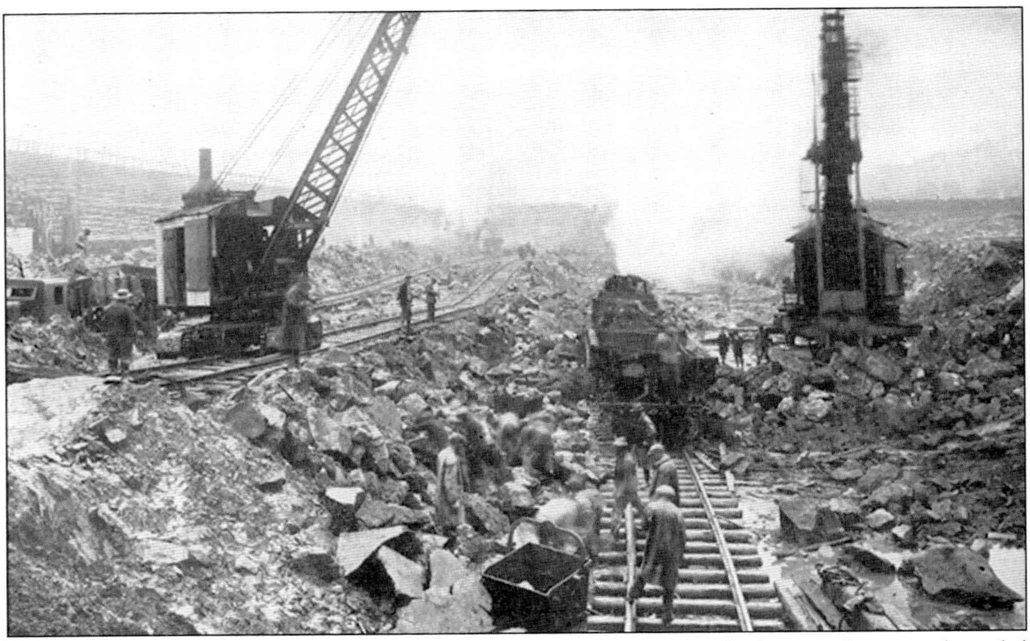

Workers used explosives to blast through the soft rock of the riverbed to get down to the solid granite on which the dam would rest. Steam shovels then cleared away the debris. Excavations for the dam, powerhouse, and tailrace were made in depths up to 40 feet. In all, nearly a half million cubic yards of Susquehanna riverbed were dug out for the dam's foundation.

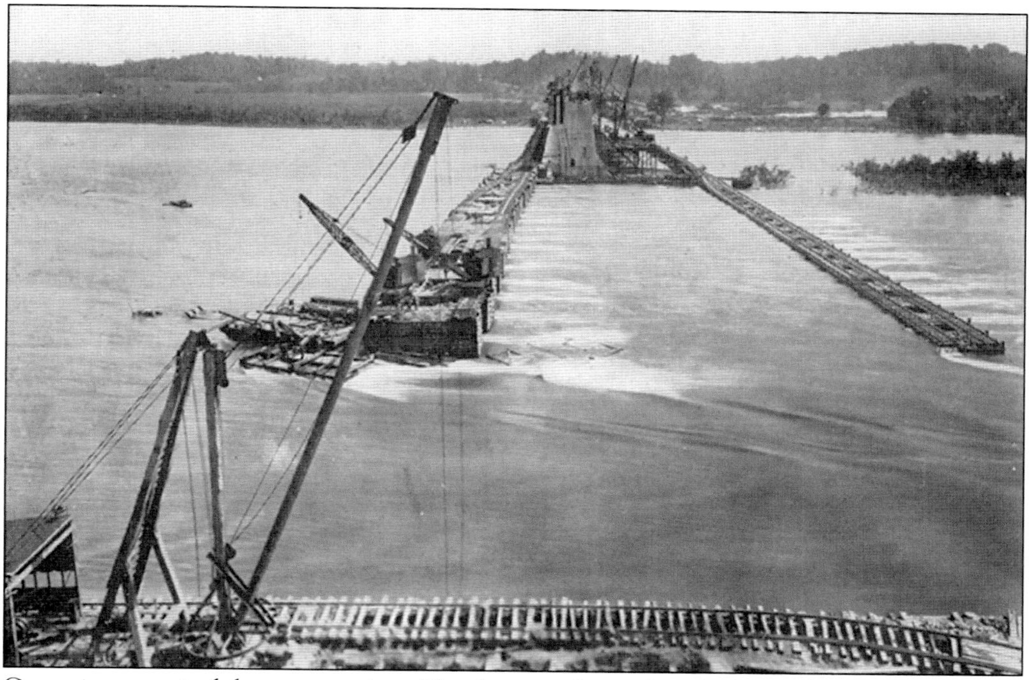

Conowingo required the construction of four huge, timber-crib cofferdams, which were filled with rock and earth to channel the Susquehanna through this 660-foot-wide gap while construction on the two shores occurred. The powerhouse cofferdam on the west bank enclosed some 14 acres of riverbed. Opposite it on the east bank, seen here, is the main closure cofferdam for the spillway.

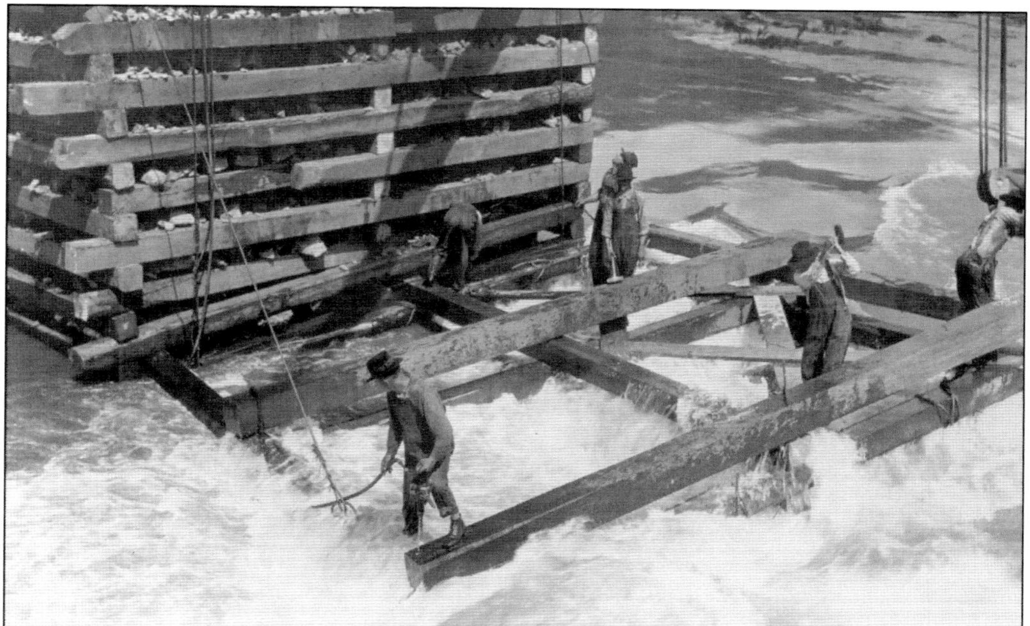

Men worked knee-deep in Susquehanna waters to frame the cofferdams, which were then filled with rock and rubble. Almost 8,000,000 feet of timber were used in the framing, and the cofferdams were constructed in merely three months.

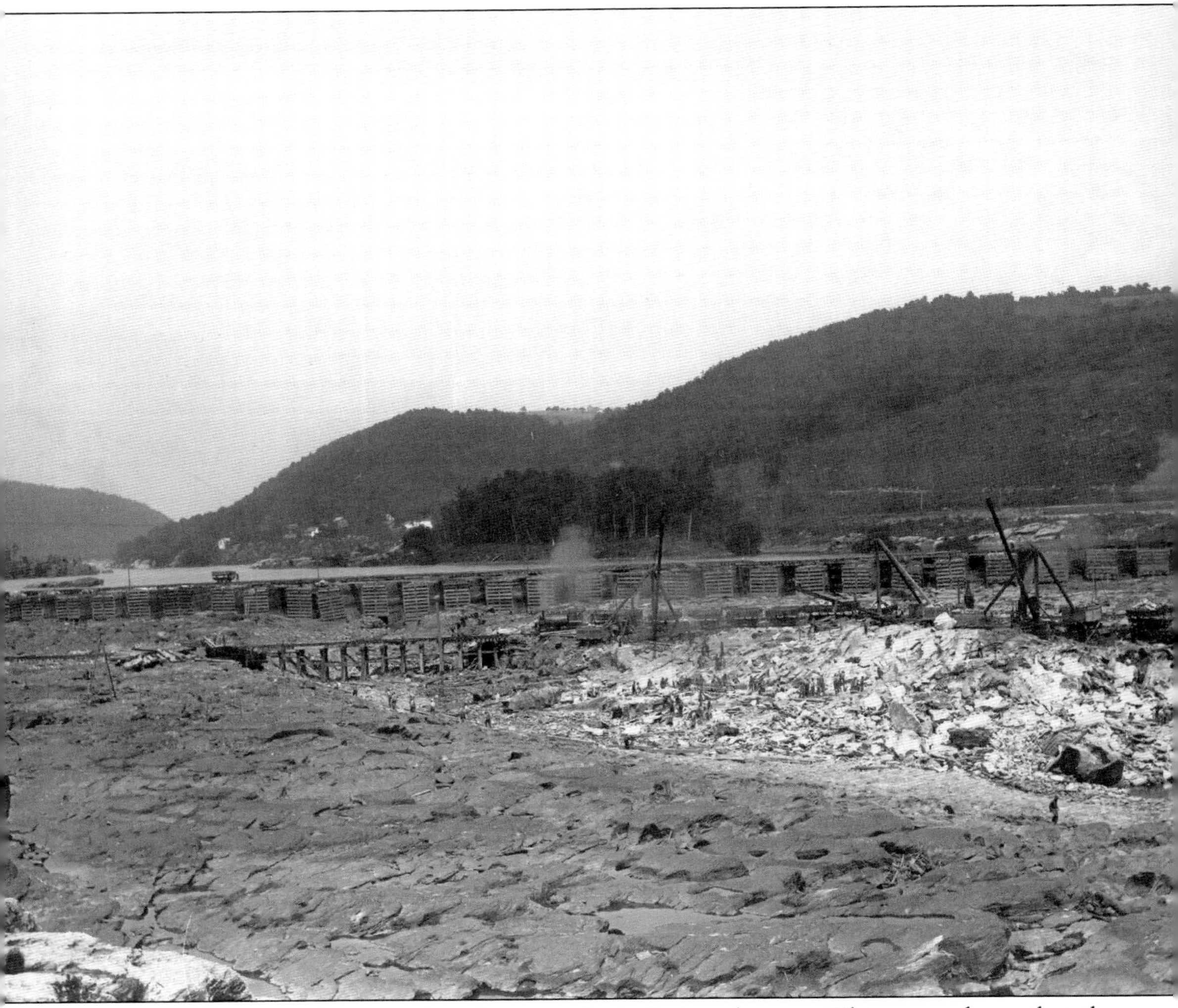

Hydro dams sit in the water, but they must be built in a dry area. This scene shows that the cofferdams not only diverted the river but also allow the enclosed area to be pumped free of water and dried out. With the river drained down to bedrock, construction workers proceeded to excavate the river bottom and erect the dam in a dry environment. In this photograph, an army of men blast, chip, and hammer the riverbed to excavate a swale, or shallow trench, which ran down the middle of the river. This depression was created to ensure that when the river flowed through the center opening of the dam, it would follow this man-made channel rather than spread out toward the shores and flood areas where the spillway and powerhouse would be erected. Running adjacent to the swale is a temporary railway to haul away the excavated stone, and a short timber bridge is seen where this train crossed over the swale. (Courtesy of the Cecil County Historical Society.)

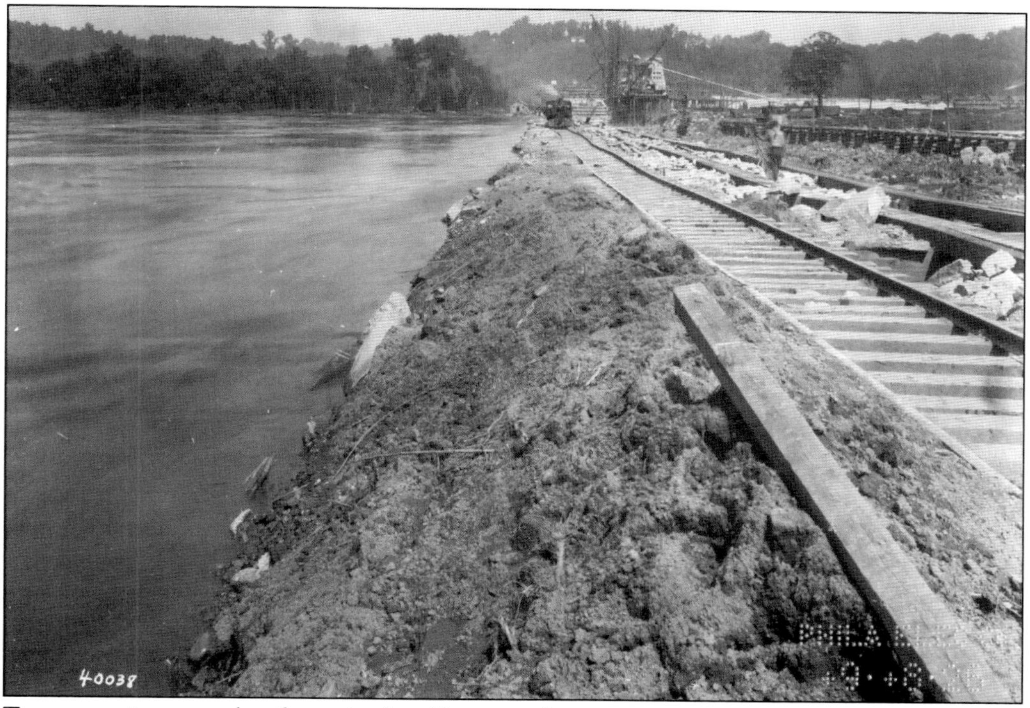

Transportation was a key factor in the efficiency of operations at Conowingo. The Columbia & Port Deposit Branch of the Pennsylvania Railroad (C&PD) already ran along the Cecil County side of the river where the spillway was built. The trains provided an excellent means of transporting supplies, while spur lines, such as seen here, were constructed to deliver materials from the main C&PD rail line to the worksite.

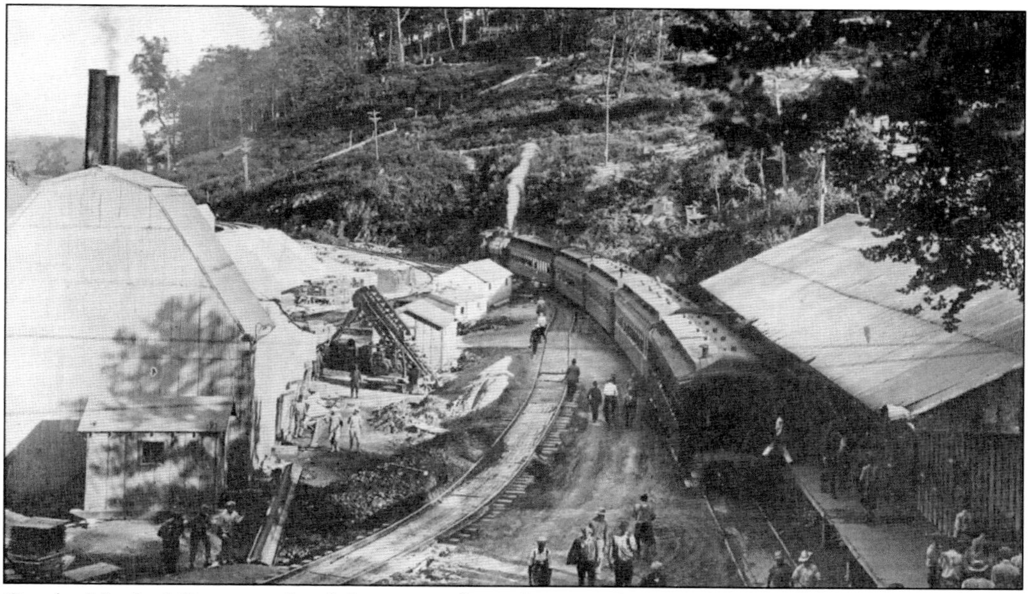

On the Harford County side of the river, where the powerhouse was constructed, a new rail line was built from Havre de Grace, Maryland, to the damsite, a distance of approximately nine miles. For most of this run, the towpath of an old colonial-era canal provided an excellent right-of-way. Here, the train is leaving the work camp on the Harford County side.

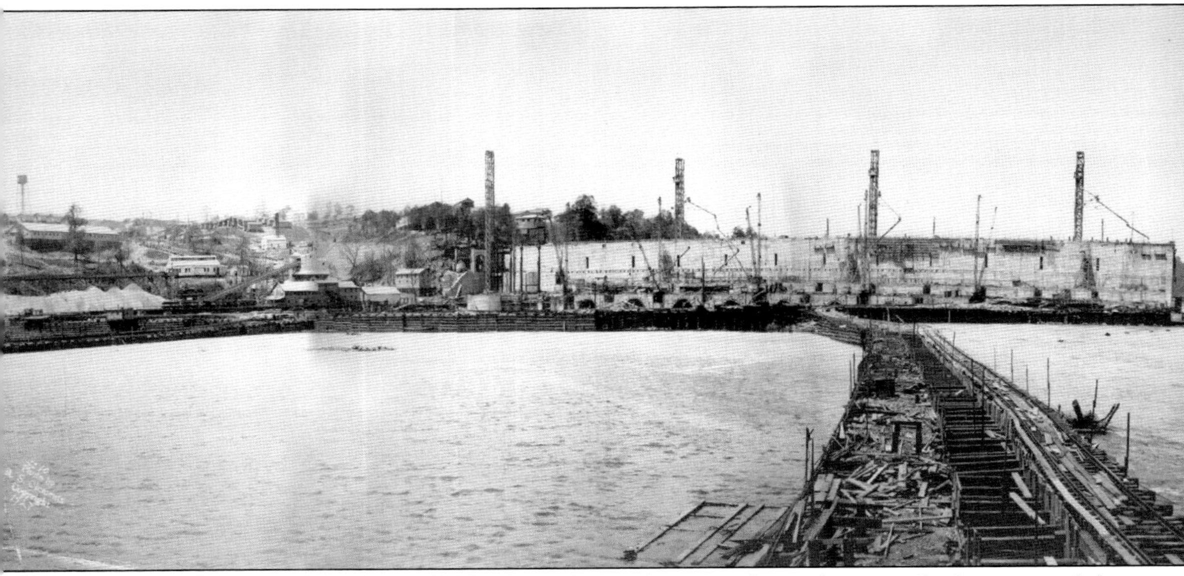

The new rail line from Havre de Grace continued on spur lines that were built on top of the cofferdams. In this way, supplies and materials coming up the Harford County side of the river were carried directly to the powerhouse. This panoramic view of the construction site shows the powerhouse building outward from the Harford County shore on the west (left) and the spillway building outward from the Cecil County shore to the east (right). The large-scale construction project brought abundant and good-paying jobs to what was then a remote region of Maryland.

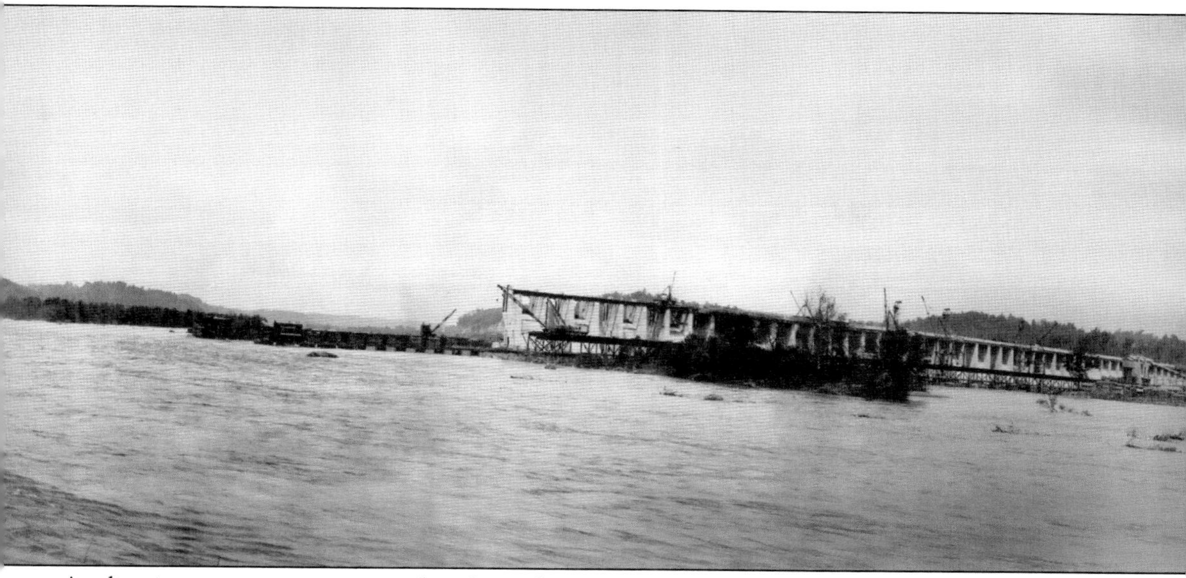

Applications came pouring in, and workers of many nationalities and ethnic backgrounds arrived in great numbers. The Boston engineering firm Stone & Webster opened a recruitment office in the village of Conowingo to hire the army of workers, and the Arundel Corporation sent them up in busloads from its offices in Baltimore. In all, about 5,500 people worked on the project, and its builders boasted that "Conowingo was built without any loss of time in labor troubles." (Courtesy of the Harford County Historical Society.)

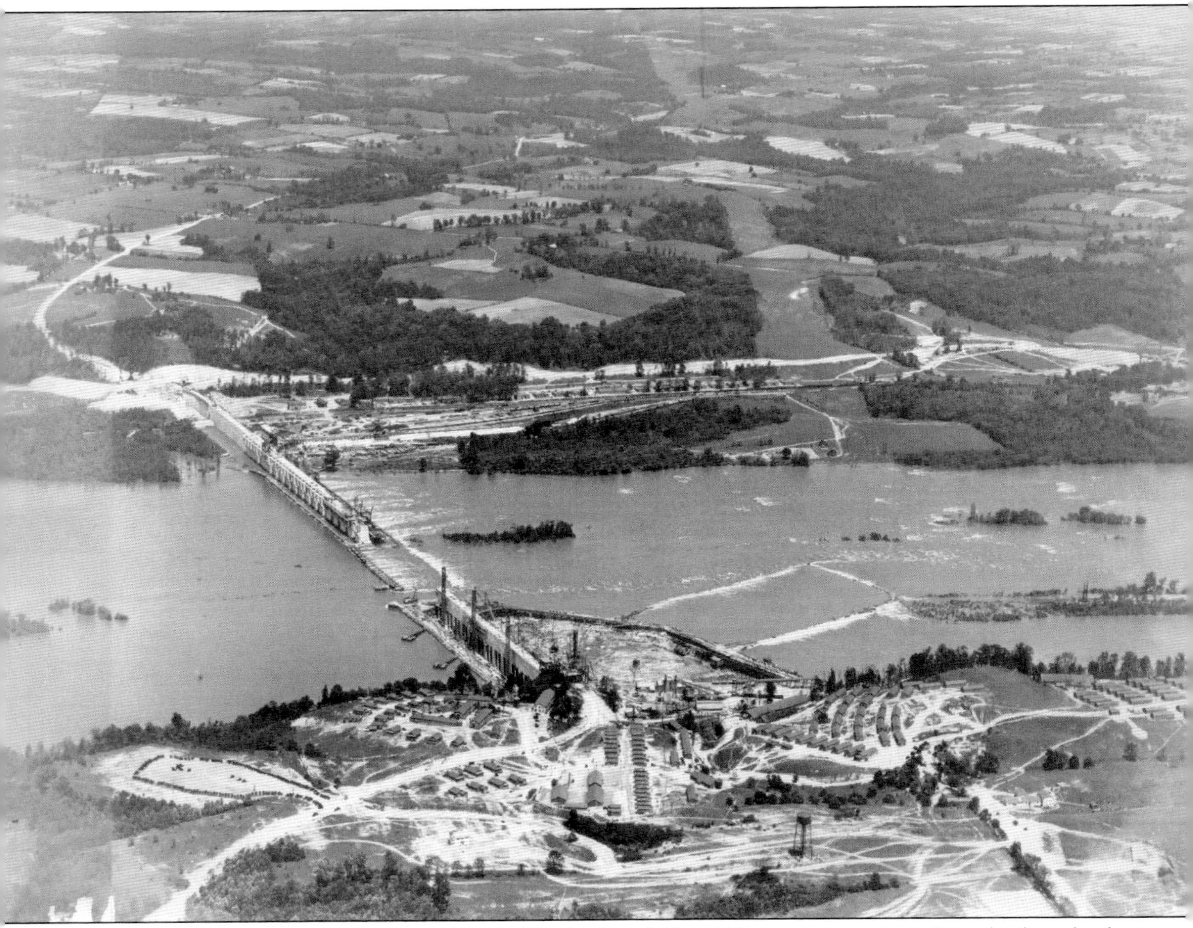

In the 1920s, the populations of both Harford and Cecil Counties numbered in the hundreds. Literally overnight, thousands of laborers, both locals and outsiders, swarmed the region. In a 1975 memoir written for the *Baltimore Sun*, Curtis Poist, a construction worker from Port Deposit, shares his recollections of the transformation of the quiet river valley. He writes, "Any able-bodied man who wanted a job could get one at 35 cents an hour for common labor, 60 cents for skilled work. In 1926, I was 19 years old. I chopped trees and grubbed out vines and saplings to clear a seven- or eight-acre area on the Harford County side. Workmen of all nationalities moved in and quickly took over every rentable room, apartment and house in Port Deposit and surrounding communities. But most of the workmen lived in scattered shantytowns built by the company." (Courtesy of the Hagley Museum.)

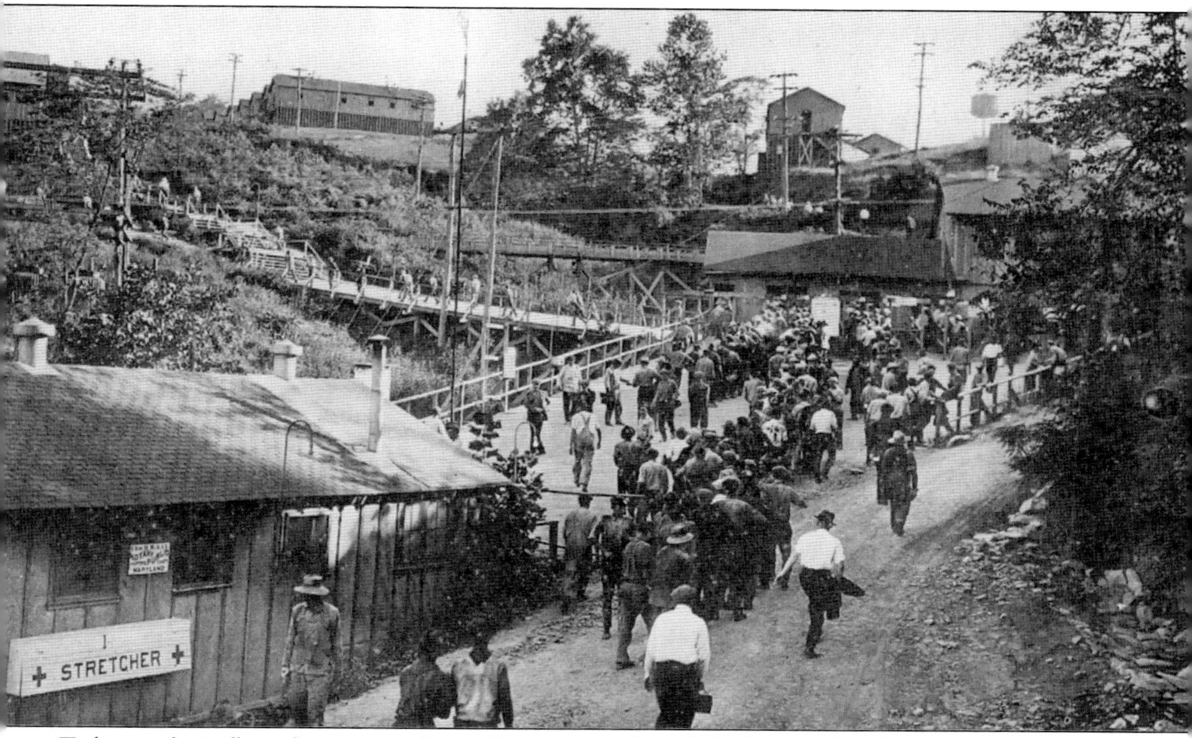

To house the influx of men, two identical camps were hastily built on each side of the river, with a mess hall, hospital, clean water supply, washhouses, and sanitation systems. Each bunkhouse housed 28 men, sleeping in single iron cots. Col. Claude B. Sweezy, a retired warden of the Maryland Penitentiary, was brought in as director of public safety to keep law and order. The workday was nine hours long, five and a half days a week, and at 3:00 p.m. on Saturdays, the job site closed down until 7:00 a.m. Monday morning. According to Curtis Poist, men were paid in cash every Saturday, and although it was the height of Prohibition, the place was "a roaring, rip-snorting boomtown hell" by nightfall, with "easy-to-find homebrew for a quarter a quart, bootleg whisky, poker and crap games running all night, and ladies of easy virtue on hand to separate a man from his money."

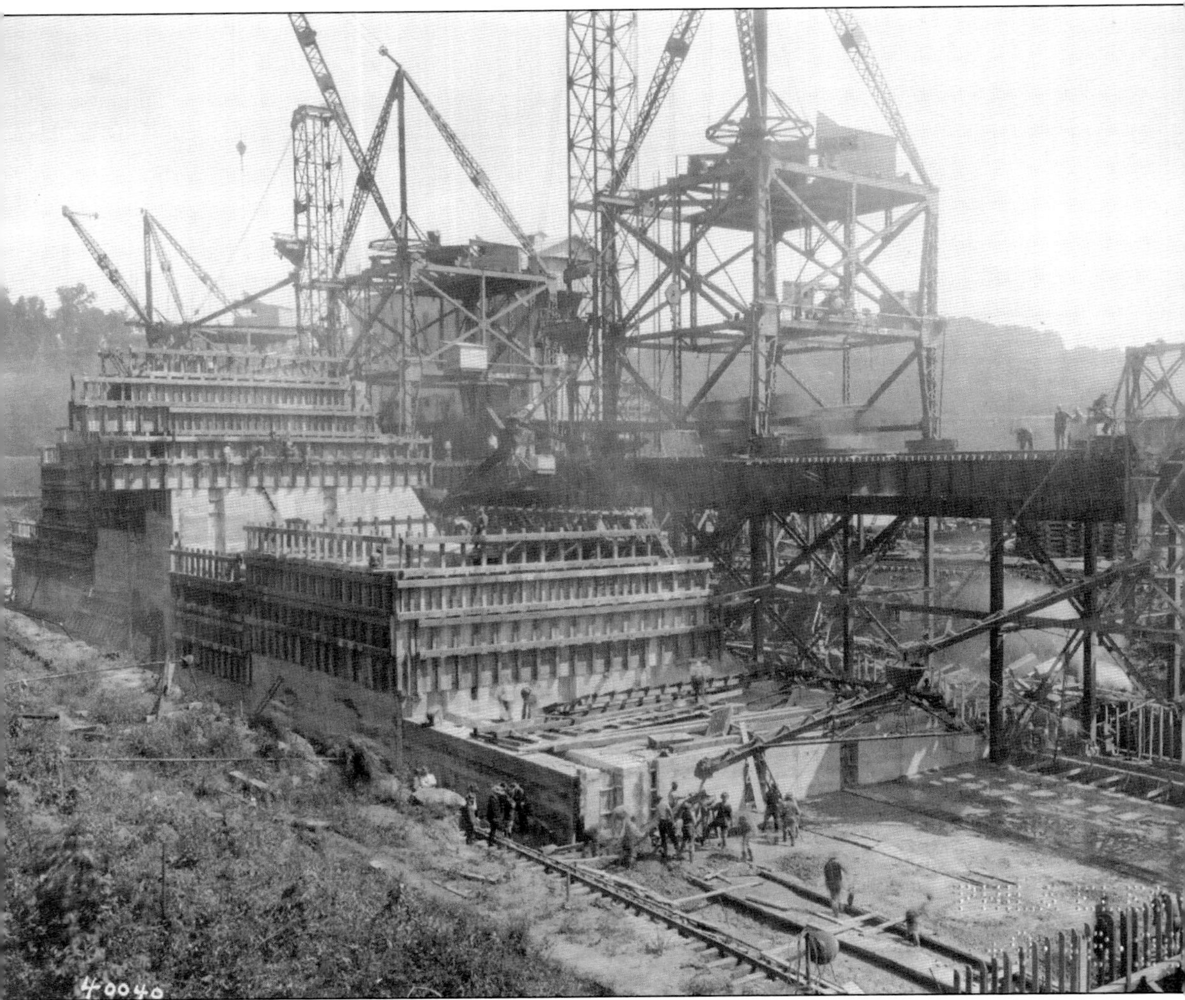

Conowingo Dam and powerhouse contains nearly 700,000 cubic yards of concrete, all of which was poured in a remarkable 17 months. The largest month's pour was 55,500 cubic yards, or a daily rate of 3,600 cubic yards. Shown here is the method devised by the construction engineers to pour such immense volumes of cement efficiently. Concrete was made on-site at two mixing plants, one on each shore, carried by gasoline dinkeys to the base of pouring towers, hoisted to the top, flowed by gravity down long articulating chutes, and, finally, poured into wooden forms in which concrete puddlers spread it evenly over the subgrade.

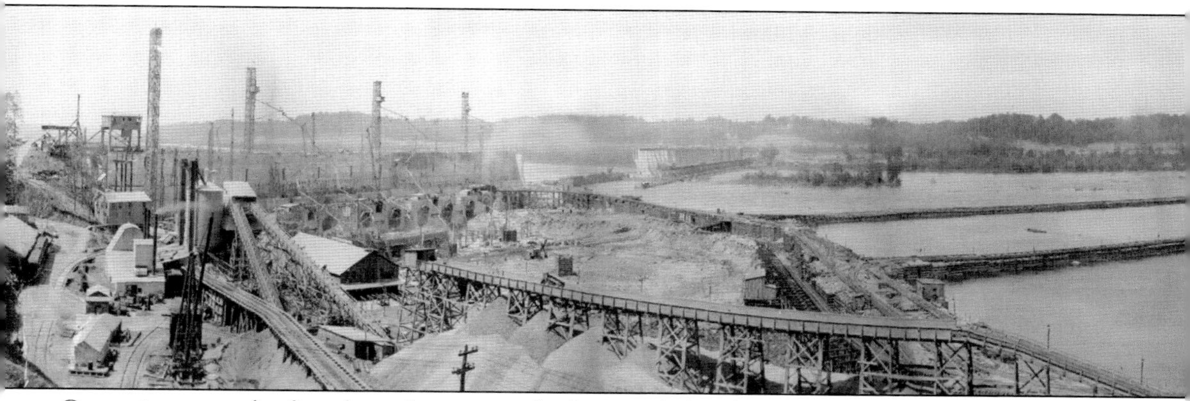

Conowingo was the first dam of major size built entirely by chuting concrete into place. The foundation for the first section of the dam was laid in the summer and fall of 1926, when the water was low. Then, in the spring of 1927, the powerhouse began to grow. This photograph shows both the powerhouse and the spillway taking shape and the river flowing through the middle of the dam. In the foreground, on the Harford County side, are the vertical hoisting towers rising above the powerhouse and the cement chutes snaking down from them. Also prominent is the concrete mixing plant where the immense amount of rock excavated for the dam's foundation was crushed and reused as the coarse aggregate for the cement that went into the first third of the concrete pours. (Courtesy of the Harford County Historical Society.)

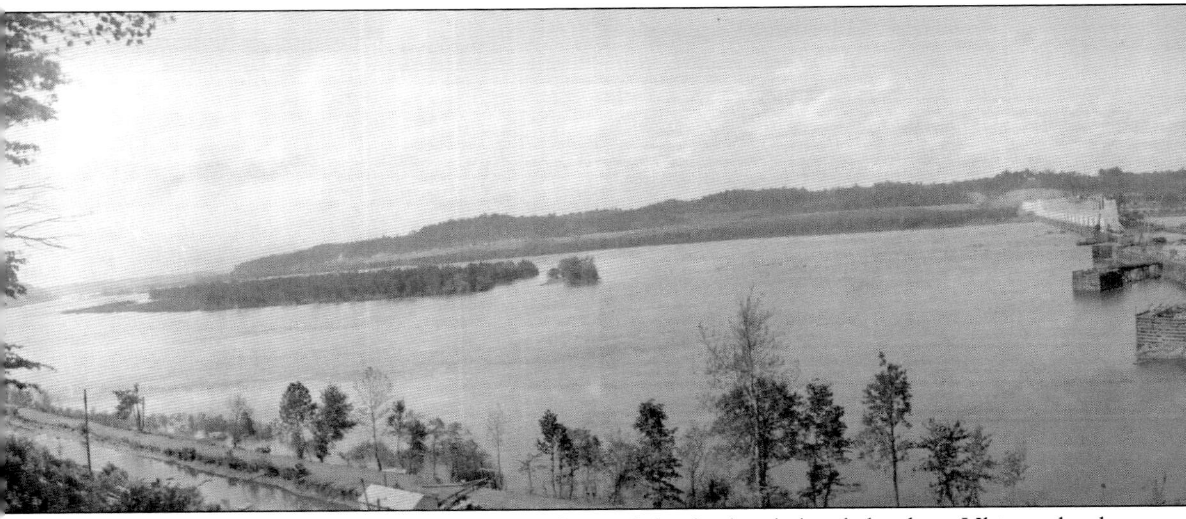

This is a view from the Harford County shore of the forebay behind the dam. Ultimately, the reservoir of water will submerge the giant square intake ports. Gravity causes the water to fall through these intakes into tubes called penstocks running from the forebay into the plant. Water then flows down the penstocks to turbine propellers, which are turned by the moving water. The shaft from each turbine goes up to a generator, which produces the power. Installed at the

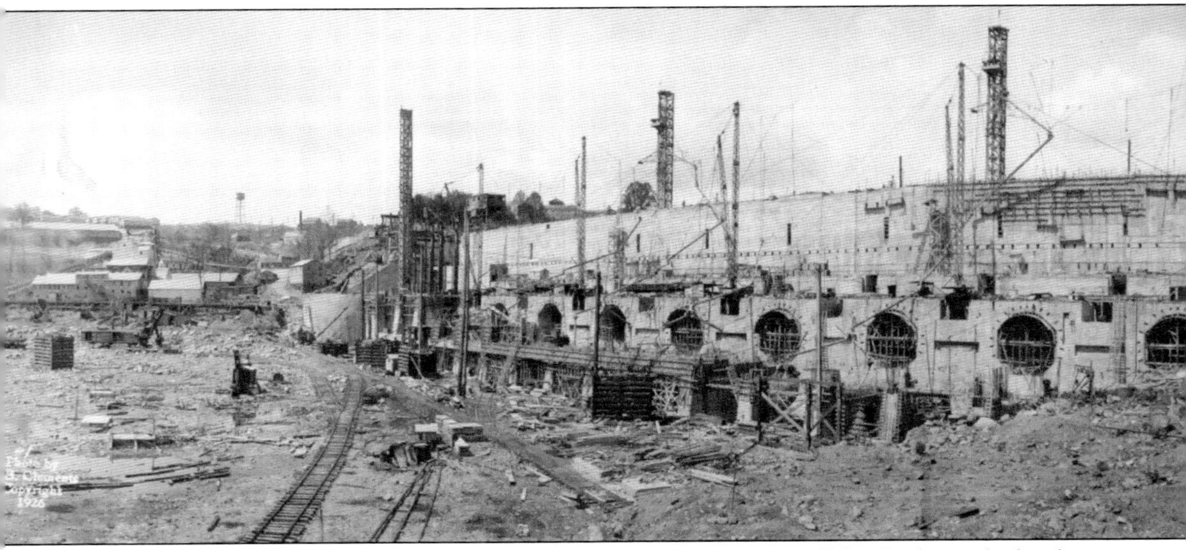

This view is of the Conowingo tailrace, which is where water will be discharged after having passed through the dam. Prominent are the huge openings for the dam's twelve penstocks, one for each of Conowingo's seven main generating units, another which is shared by the plant's two smaller "house" generators, and four more for future expansion. When power is generated, water will flow from the forebay through the penstocks into scroll cases, where each turbine is

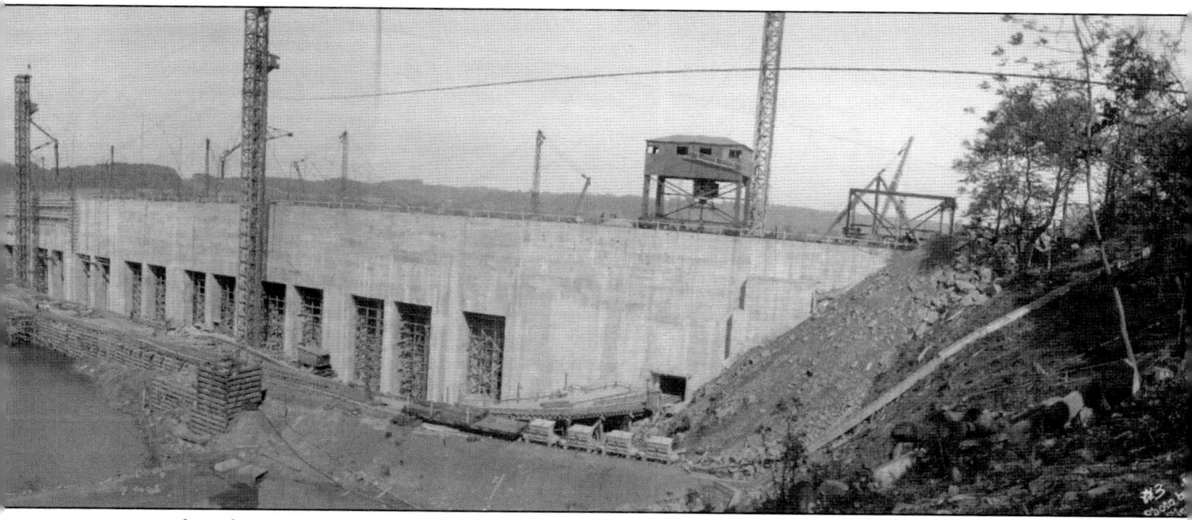

entrance of each penstock are trash racks, metal grates that prevent waterborne debris from causing destruction of the penstock, shutoff valve, waterwheel, or draft tube. The powerhouse cofferdam is seen in front of the intakes, and an ingenious tunnel temporarily cuts through the dam to allow the train to bring materials to this side of the dam. (Courtesy of the Harford County Historical Society.)

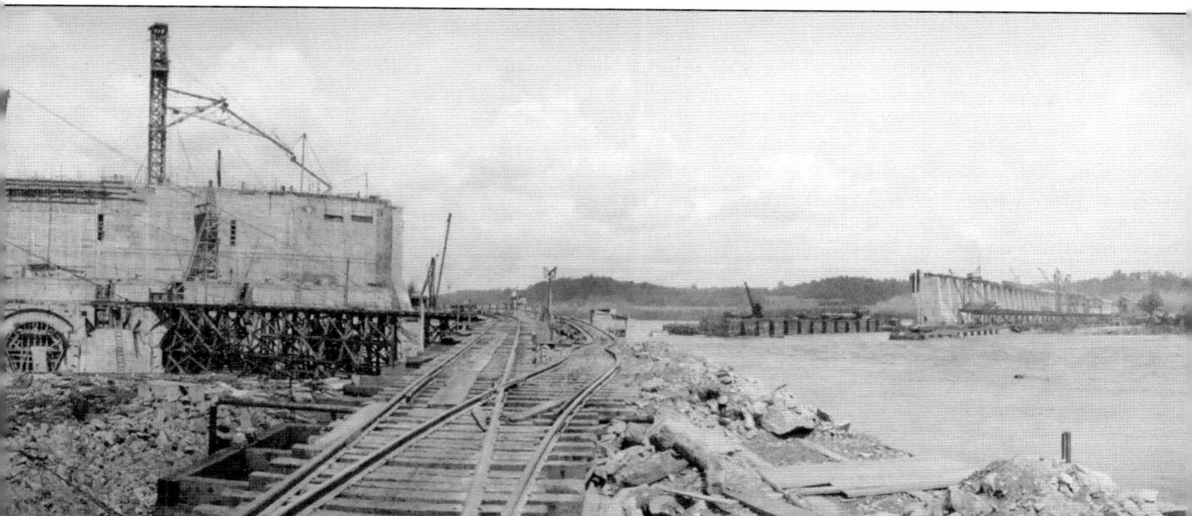

housed; it then exits the tailrace through what are called draft tubes. In this photograph, both the scroll cases and draft tubes are yet to be installed. At the far left, steelworkers are beginning to erect the superstructure of the dam's vast, seven-story, 620-foot room known as turbine hall, which housed all of Conowingo's mammoth turbines and generating equipment. (Courtesy of the Harford County Historical Society.)

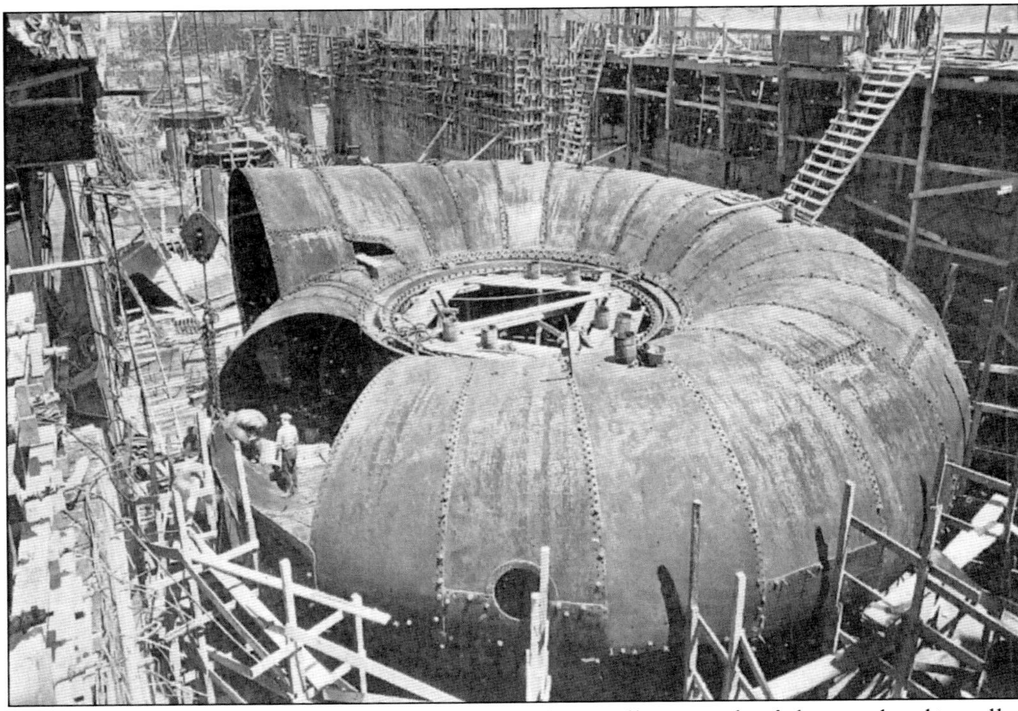

Each of the seven generating units was equipped with a scroll case made of plate steel and installed in the heart of the powerhouse. The scroll case is situated between the penstock, where water enters the dam, and the draft tube, where water exits. As water rushes from the forebay through the penstock, it hits the scroll case. The scroll case coils a full 360 degrees around the turbine to direct the water in a clockwise spiral. It also gradually narrows, thereby directing the water to strike the spinning turbine blades at the proper angle and with uniform force greater than 150 tons of water per second. The scroll case seen here has a large opening in its top where the waterwheel and shaft will be dropped. A small hatch on the side allows access for future maintenance and repairs. Ultimately, each scroll case is buried in solid concrete, thus permanently securing it within the dam.

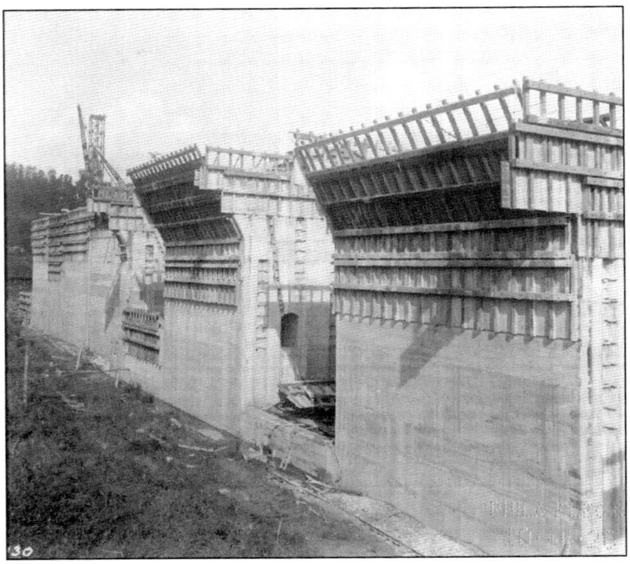

The crest of Conowingo's spillway, seen here, was built to be 86 feet above the base of the dam. The spillway was poured in alternating 52-foot sections with 38-foot notches left low to allow the river to handle periodic flood flows. In this way, work could be carried on continuously regardless of unusually high water.

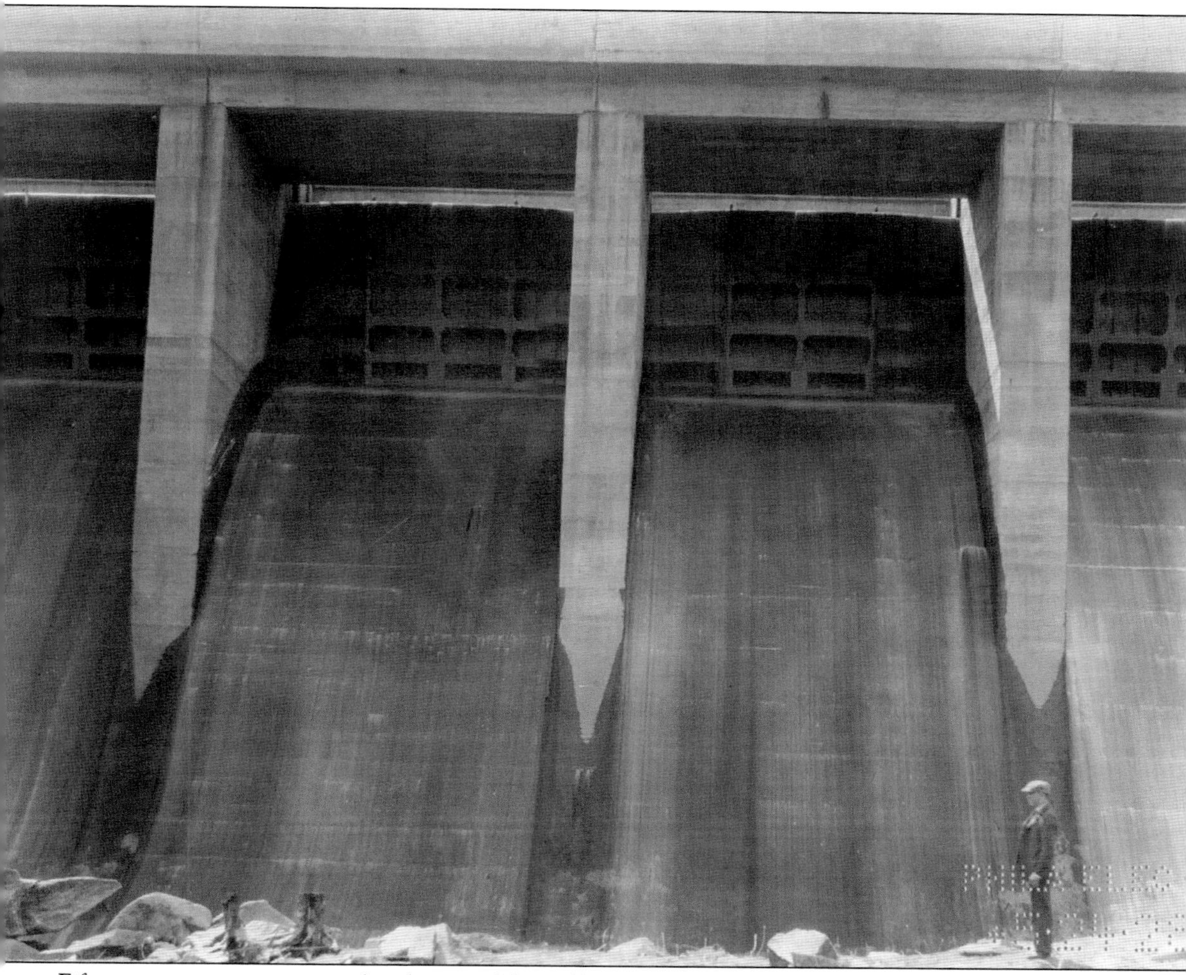

Fifty crest gates were mounted at the top of the spillway, each measuring 22.5 feet high, 41 feet wide, and weighing 42 tons. Ordinarily, the dam holds back the reservoir so that water flows through the dam's generators. But during high-water periods, when the river flow exceeds generating capacity, the crest gates are opened to spill excess water over the top of the dam. The man standing on the Susquehanna riverbed gives an indication of the soaring height of the spillway.

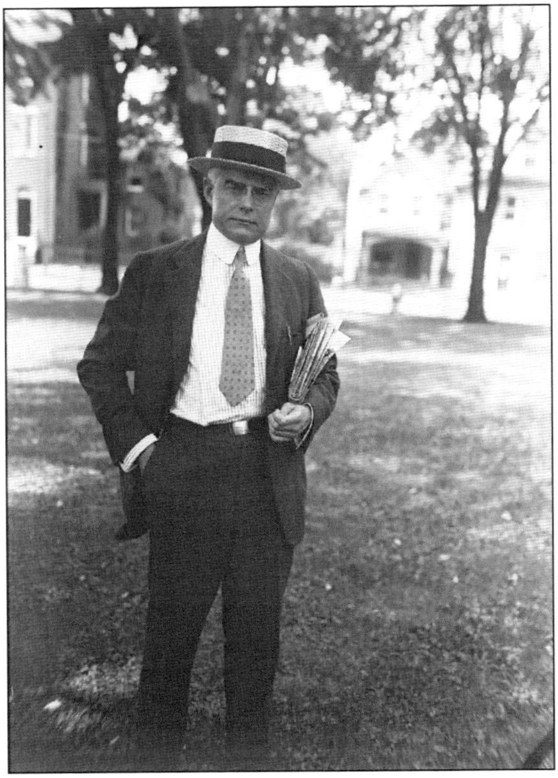

With construction well under way, Conowingo became the central issue of Maryland's 1926 gubernatorial election, when sitting governor Albert C. Ritchie (left) squared off against challenger Addison E. Mullikin (below), who was dismayed that Philadelphia would reap the benefits of this hydro dam built on a Maryland river. Mullikin brushed aside the fact that Baltimore was already receiving all the power it needed from an existing generating facility, Holtwood Dam, so there was neither the demand nor financing for more electricity in Maryland. Meanwhile, Philadelphians needed extra power to run a growing network of electrified streetcars, and private backers were able to muster the financing for the massive project. In the weeks before the election, Mullikin stirred Conowingo into a full-blown controversy. (Both, courtesy of the Library of Congress.)

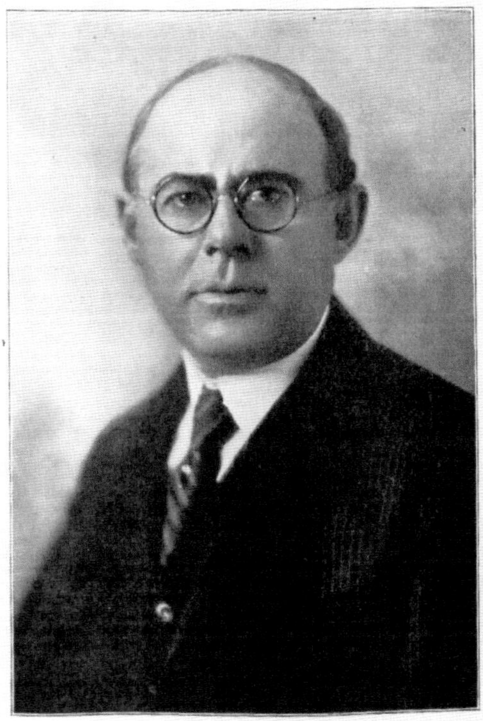

ADDISON E. MULLIKIN

Mullikin accused Ritchie of being beholden to Wall Street financiers and of receiving kickbacks from wealthy capitalists. Even worse was Mullikin's refrain, satirized by newspaper cartoonists, that Ritchie had sold out the birthright of every Maryland citizen—the right to exploit its natural resources for its sole benefit. (Courtesy of the *Baltimore Sun*.)

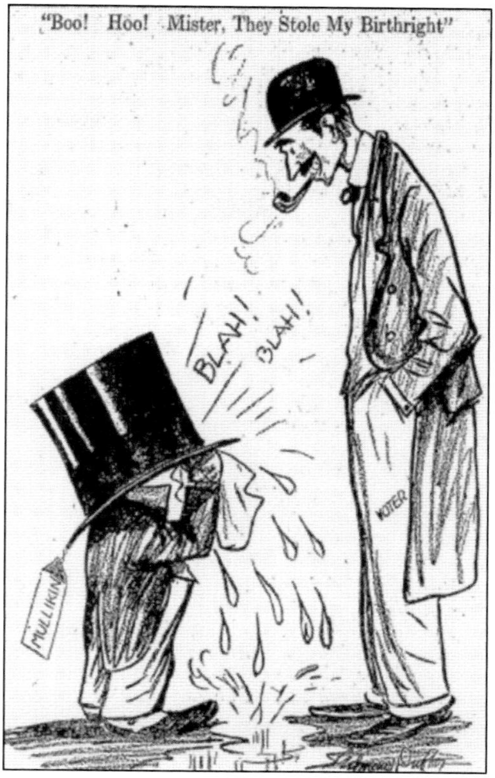

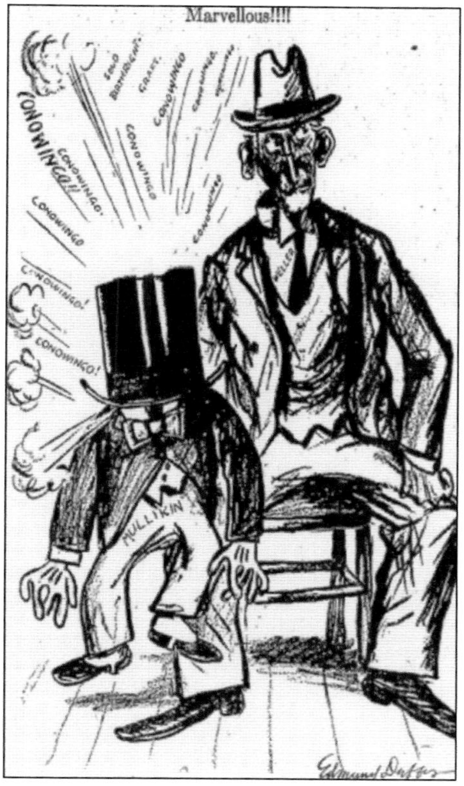

A political boycott against Governor Ritchie was organized by the Baltimore Federation of Labor, which claimed, "Sale of the power in Pennsylvania is objected to . . . on the ground that it will result ultimately in the sale of power generated in Maryland to Pennsylvania cheaper than power can be bought in Baltimore." Emboldened by the union's boycott threat, Mullikin relentlessly hammered the issue. (Courtesy of the *Baltimore Sun*.)

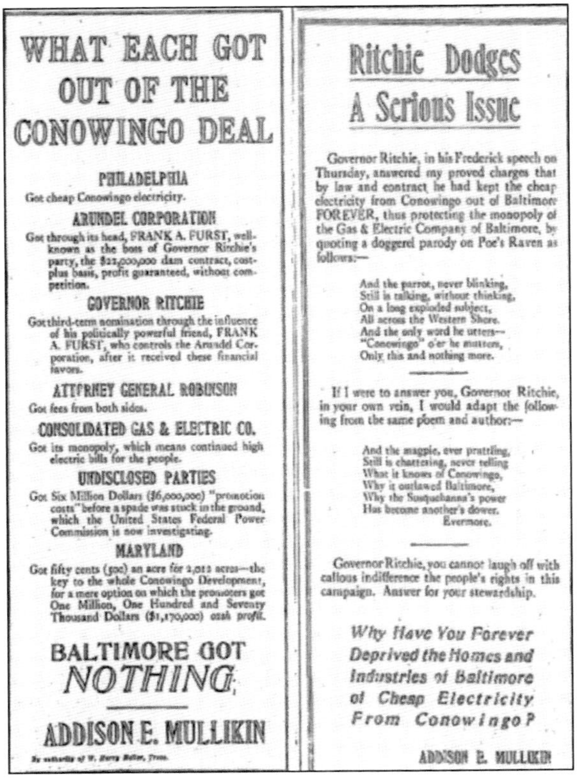

WHAT EACH GOT OUT OF THE CONOWINGO DEAL

PHILADELPHIA

Got cheap Conowingo electricity.

ARUNDEL CORPORATION

Got through its head, FRANK A. FURST, well-known as the boss of Governor Ritchie's party, the $22,000,000 dam contract, cost-plus basis, profit guaranteed, without competition.

GOVERNOR RITCHIE

Got third-term nomination through the influence of his politically powerful friend, FRANK A. FURST, who controls the Arundel Corporation, after it received these financial favors.

ATTORNEY GENERAL ROBINSON

Got fees from both sides.

CONSOLIDATED GAS & ELECTRIC CO.

Got its monopoly, which means continued high electric bills for the people.

UNDISCLOSED PARTIES

Got Six Million Dollars ($6,000,000) "promotion costs" before a spade was stuck in the ground, which the United States Federal Power Commission is now investigating.

MARYLAND

Got fifty cents (50c) an acre for 2,012 acres—the key to the whole Conowingo Development, for a mere option on which the promoters got One Million, One Hundred and Seventy Thousand Dollars ($1,170,000) cash profit.

BALTIMORE GOT *NOTHING*

ADDISON E. MULLIKIN

By authority of W. Harry Weller, Treas.

Ritchie Dodges A Serious Issue

Governor Ritchie, in his Frederick speech on Thursday, answered my proved charges that by law and contract he had kept the cheap electricity from Conowingo out of Baltimore FOREVER, thus protecting the monopoly of the Gas & Electric Company of Baltimore, by quoting a doggerel parody on Poe's Raven as follows:—

And the parrot, never blinking,
Still is talking, without thinking,
On a long exploded subject,
All across the Western Shore.
And the only word he utters—
"Conowingo" o'er he mutters,
Only this and nothing more.

If I were to answer you, Governor Ritchie, in your own vein, I would adapt the following from the same poem and author:—

And the magpie, ever prattling,
Still is chattering, never telling
What it knows of Conowingo,
Why it outlawed Baltimore,
Why the Susquehanna's power
Has become another's dower.
Evermore.

Governor Ritchie, you cannot laugh off with callous indifference the people's rights in this campaign. Answer for your stewardship.

Why Have You Forever Deprived the Homes and Industries of Baltimore of Cheap Electricity From Conowingo?

ADDISON E. MULLIKIN

The battle culminated in these advertisements placed days before the election. In them, Mullikin insisted that "Baltimore Got Nothing" from the deal and asked Governor Ritchie, "Why have you forever deprived the homes and industries of Baltimore of cheap electricity from Conowingo?" Ritchie issued a novel rebuttal: he defended his support for the project in an elegant parody of Baltimore's famous man of letters, Edgar Allan Poe. (Courtesy of the *Baltimore Sun.*)

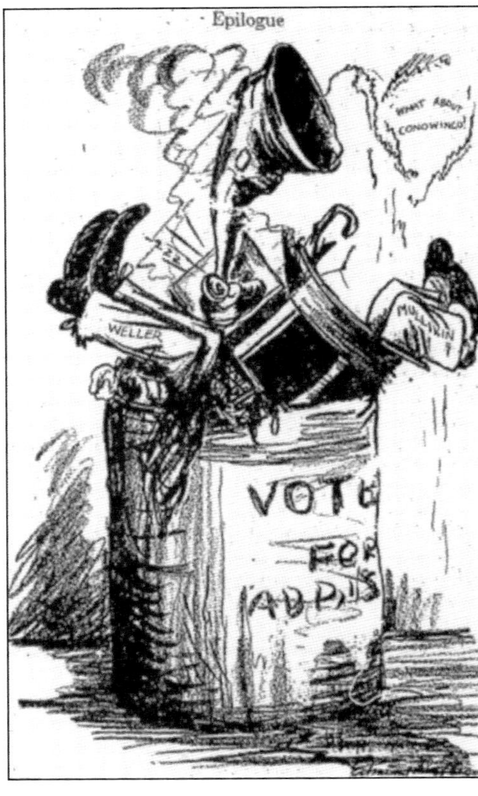

On election day, both Mullikin and the Conowingo controversy were tossed into the trash can of history. Governor Ritchie prevailed, and his vision of harnessing the Susquehanna for the region was realized. "I think we might as well say goodbye to the Conowingo issue," he said, "There never was anything to it except propaganda." (Courtesy of the *Baltimore Sun.*)

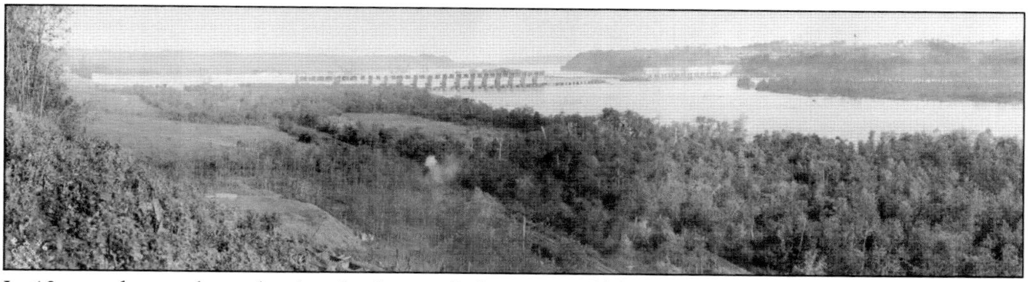

In 12 months, work on the dam had moved along incredibly well. In this view looking downriver from the Cecil County shore, the powerhouse has risen on the far Harford County riverbank and the spillway is extending out from the Cecil County bank to meet it. Visible are the alternating spillway sections and the center opening of the dam, which allowed the river to flow through it. Note in the foreground the white puffs of a steam engine chugging along the Columbia & Port Deposit tracks running through Susquehanna bottomland. In the 19th and early 20th centuries, the 34-mile-long railroad was the main line between Columbia, Pennsylvania, and Port Deposit, Maryland. Its low-lying right-of-way was so near the riverbank that it would become submerged below pond level when the dam opened, so it had to be relocated to higher ground. (Courtesy of the Harford County Historical Society.)

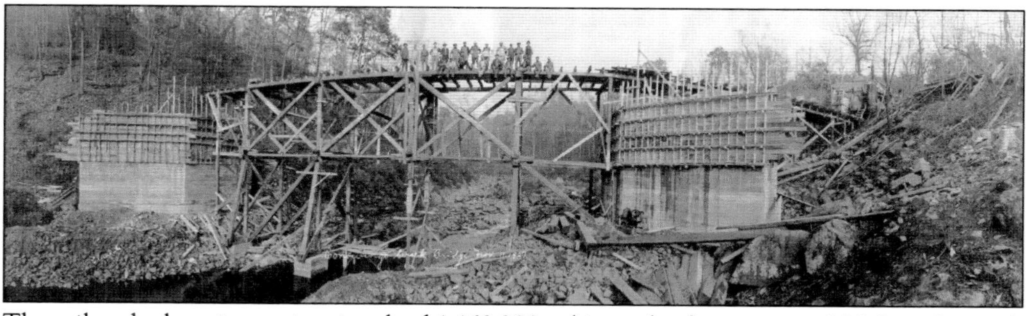

The railroad relocation project involved 1,460,000 cubic yards of excavation, 900 feet of tunnel, 42,000 cubic yards of concrete for bridges and culverts, and the laying of 20 miles of new track. In this photograph, the construction crew is taking a break from the construction of a new bridge over Conowingo Creek. (Courtesy of the Harford County Historical Society.)

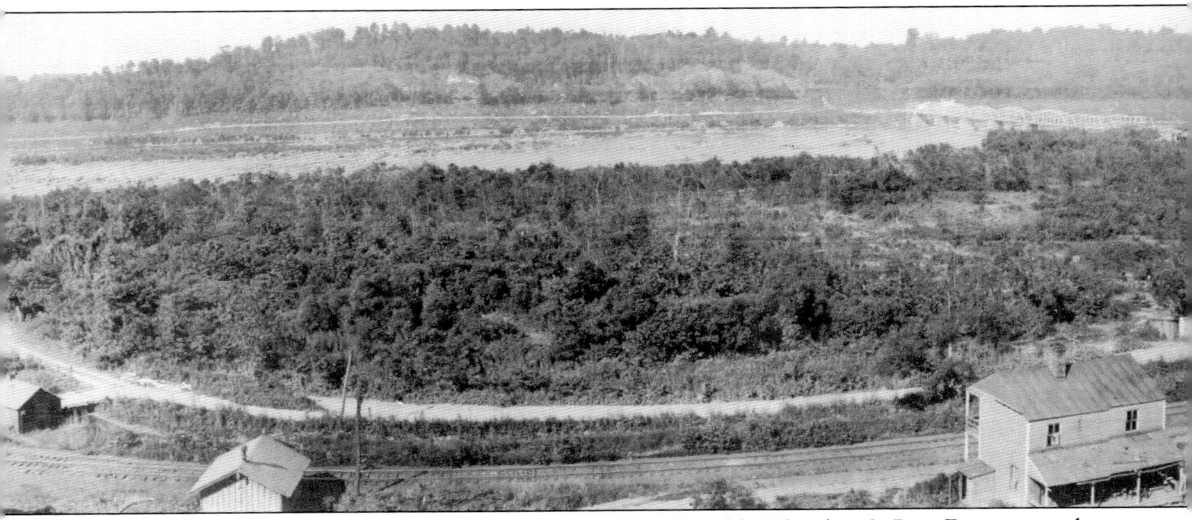

This photograph of the village of Conowingo shows the old Columbia & Port Deposit tracks as they ran through the heart of town, past wooden homes, shops, churches, mills, and canneries. The new railroad bridge over Conowingo Creek is seen near completion. The bridge was a harbinger,

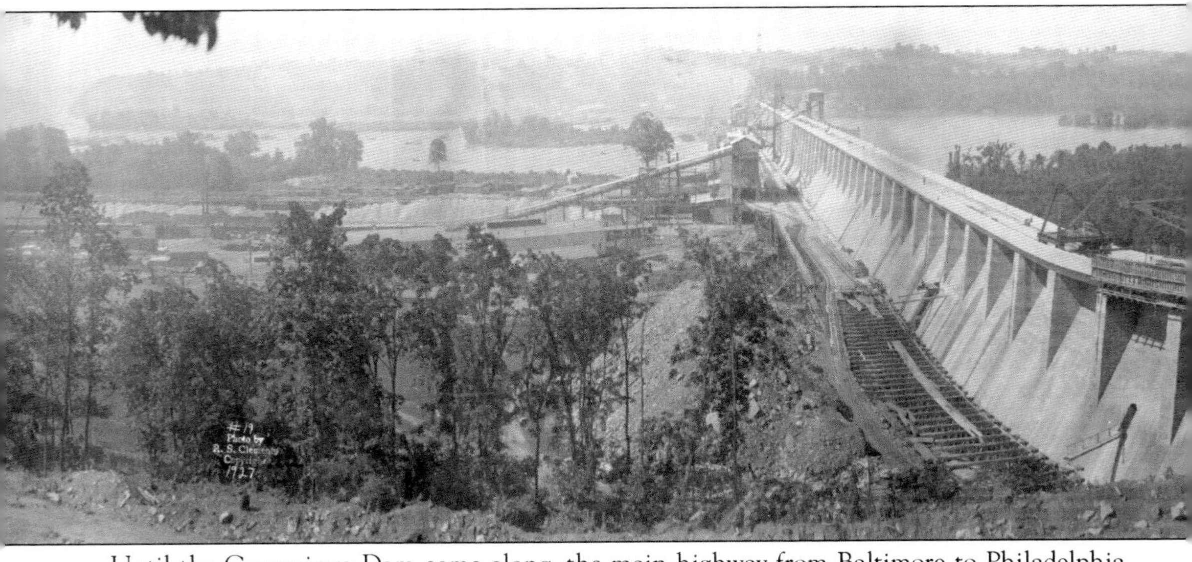

Until the Conowingo Dam came along, the main highway from Baltimore to Philadelphia, known as the Baltimore Pike, crossed the Susquehanna River two miles from the damsite, over the low-slung bridge at the village of Conowingo. That old bridge sat well below the level of the eventual pond and its relocation was necessary, thus the Conowingo Dam was also designed as a highway bridge to carry traffic on its crest. In this view from the Cecil County side of the

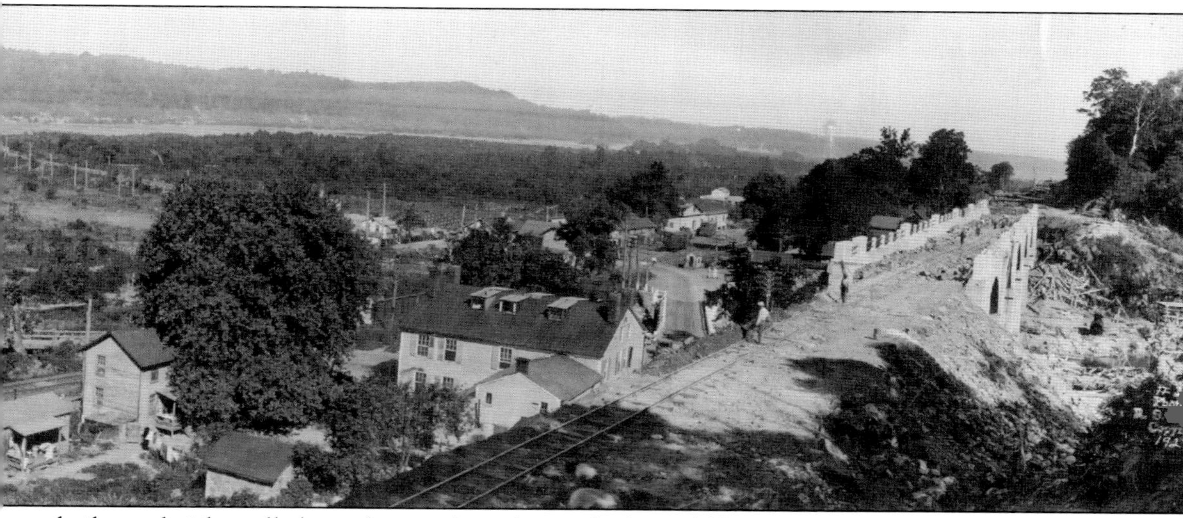

built at a height well above the town's rooftops, which would be underwater within a matter of months. (Courtesy of the Harford County Historical Society.)

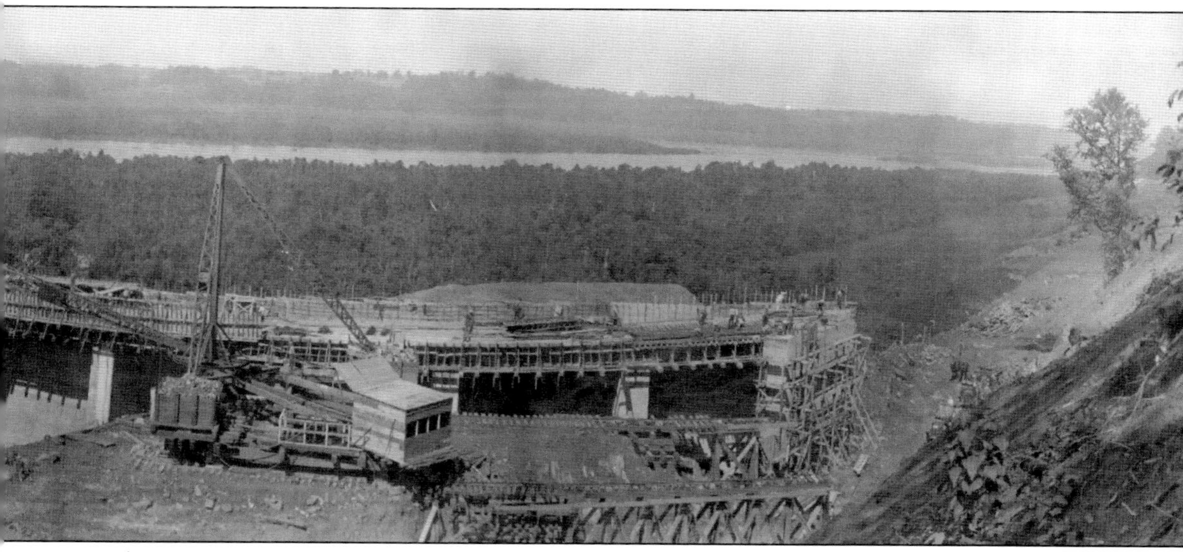

river, the approach road and the roadway over on top of the dam are nearing completion. The relocation of the river crossing to Conowingo Dam was welcomed because it shortened the trip through the region by roughly four miles, and drivers could navigate the straightened new road more safely than the hilly grades and sharp curves of the old one. (Courtesy of the Harford County Historical Society.)

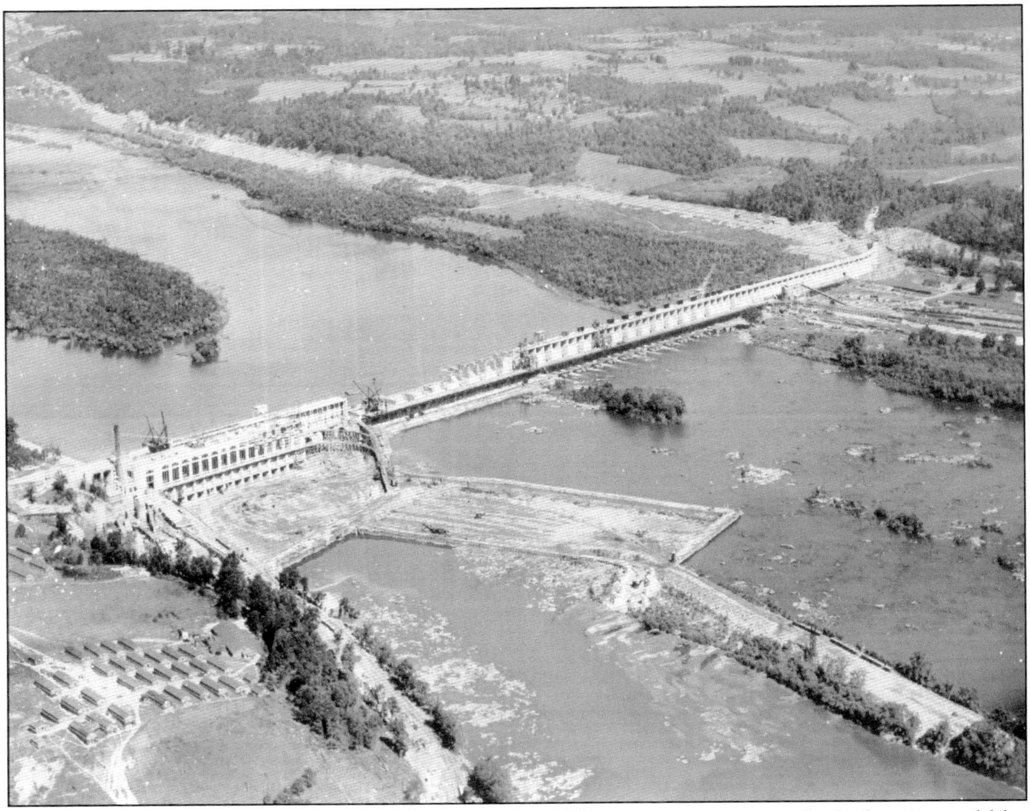

The peak of construction was reached in August 1927, a mere seven months before it would be generating power, when 3,725 men were working on the powerhouse and spillway alone. There were also 225 men working on transmission lines to Philadelphia, and 1,400 working on the railroad relocation and rerouting of the highway. (Courtesy of the Hagley Museum.)

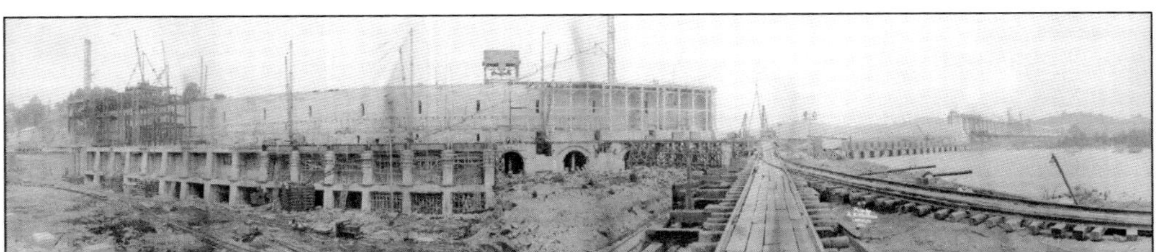

The powerhouse superstructure required some 4,566 tons of structural steel. It is supported by two rows of buttresses. Water exits the tailrace from the gaps between the lower row of buttresses. As water leaves the turbine, it makes a 30-foot vertical drop into draft tubes, which swoop in an upward curve out of the dam, a design intended to dissipate the energy of the water as it leaves the dam. (Courtesy of the Harford County Historical Society.)

Three

Susquehanna River Ice Floes and Floods

Generations of Marylanders in towns like Port Deposit, Havre de Grace, and Perryville knew the Susquehanna was capricious. During the dog days of summer, it appeared to be little more than a trickle, but a tropical storm could turn it into a torrent that surmounted the banks and ravaged nearby homes. Springtime thaws, called freshets, were especially fearsome. In winter, the Susquehanna routinely froze into thick sheets of ice; in early spring, the ice sheets thawed, broke loose, and floated down toward the Chesapeake Bay. Great blocks of ice drifted downriver and jammed together, one ice floe ramming into another and another, then froze again into gigantic dams that stretched from shore to shore.

Snowmelt coming down the watershed had nowhere to go, and it backed up behind the ice dams, overflowed the banks, and inundated low-lying towns. Ice gorges of staggering size and power had caused destruction along the river for centuries, and Marylanders had tried many inventive means to break them up. In 1920, with ice dangerously piling up in the river, Lt. Rene R. Studler was called to arms and, taking off from the airfield at Aberdeen Proving Ground, proceeded to "bomb the blue blazes out of the frozen Susquehanna."

The Aero Club of America, a group of early aviation enthusiasts, notes the event in volume 9 of the magazine *Flying* as follows: "The ice gorge in Port Deposit showed signs of loosening under the combined attack of warmer atmosphere and bombing from the aeroplane. . . . Lt. Studler dropped several 500-pound TNT bombs into the heart of the jam . . . a great agitation was visible below the jam where the aviator broke up clean ice with 220-pound bombs." Studler's run was exciting but did not succeed in loosening the blockade. His bombs broke the ice apart, but it soon backed up again against islands and bridge piers, and water pushed into the town of Port Deposit as it usually did.

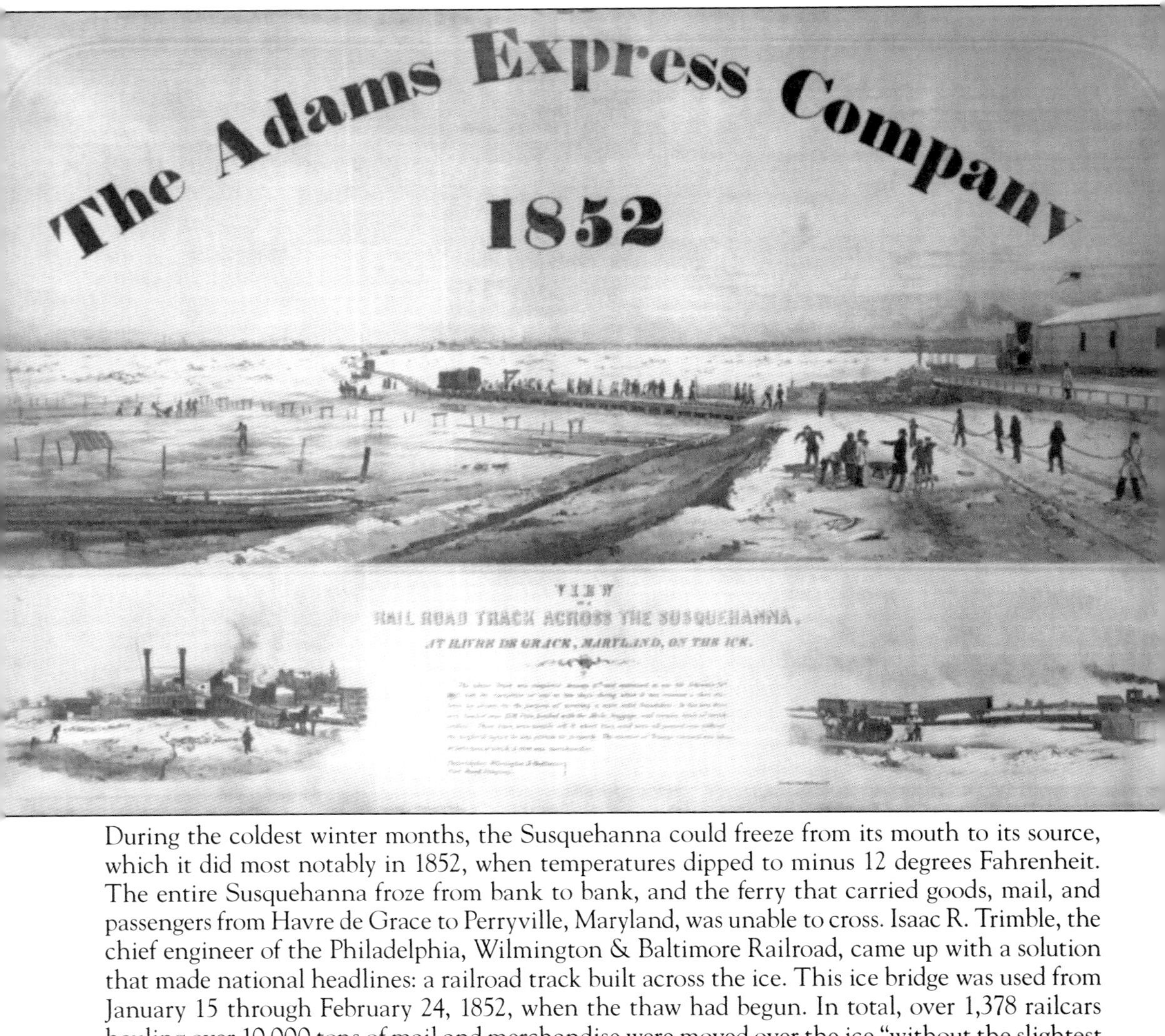

During the coldest winter months, the Susquehanna could freeze from its mouth to its source, which it did most notably in 1852, when temperatures dipped to minus 12 degrees Fahrenheit. The entire Susquehanna froze from bank to bank, and the ferry that carried goods, mail, and passengers from Havre de Grace to Perryville, Maryland, was unable to cross. Isaac R. Trimble, the chief engineer of the Philadelphia, Wilmington & Baltimore Railroad, came up with a solution that made national headlines: a railroad track built across the ice. This ice bridge was used from January 15 through February 24, 1852, when the thaw had begun. In total, over 1,378 railcars hauling over 10,000 tons of mail and merchandise were moved over the ice "without the slightest injury to any person or property." This lithograph commemorating Trimble's achievement was produced by Thomas S. Sinclair of Philadelphia, based on a drawing by F.F. Schell. The Adams Express Company, a prominent courier business of the 19th century, distributed this imprint as an advertisement to its customers. (Courtesy of the Enoch Pratt Free Library.)

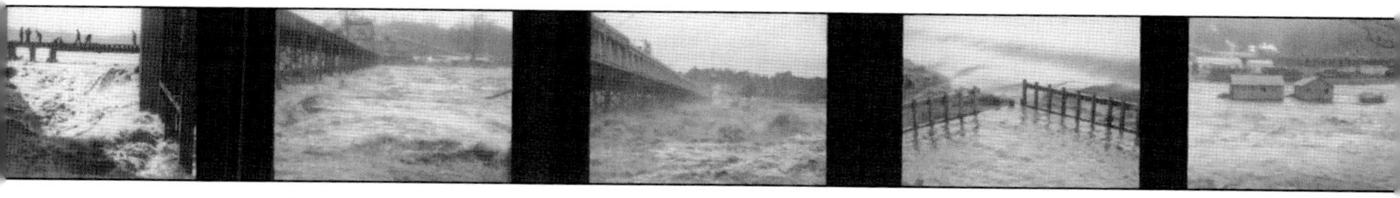

In November 1926, floods came again to the Susquehanna River, and the river rose to a record flow, threatening the Conowingo construction project. Frames from a newsreel film taken at the time dramatize the peril as the film's inter-titles ponder the whims of nature: "It is a tradition among construction men said that the surest way to bring about a great flood in a river is to start building a dam across it! The Susquehanna is no exception – for when the work is well along, reports come from the headwaters that the greatest November flood in more than 35 years is in the making."

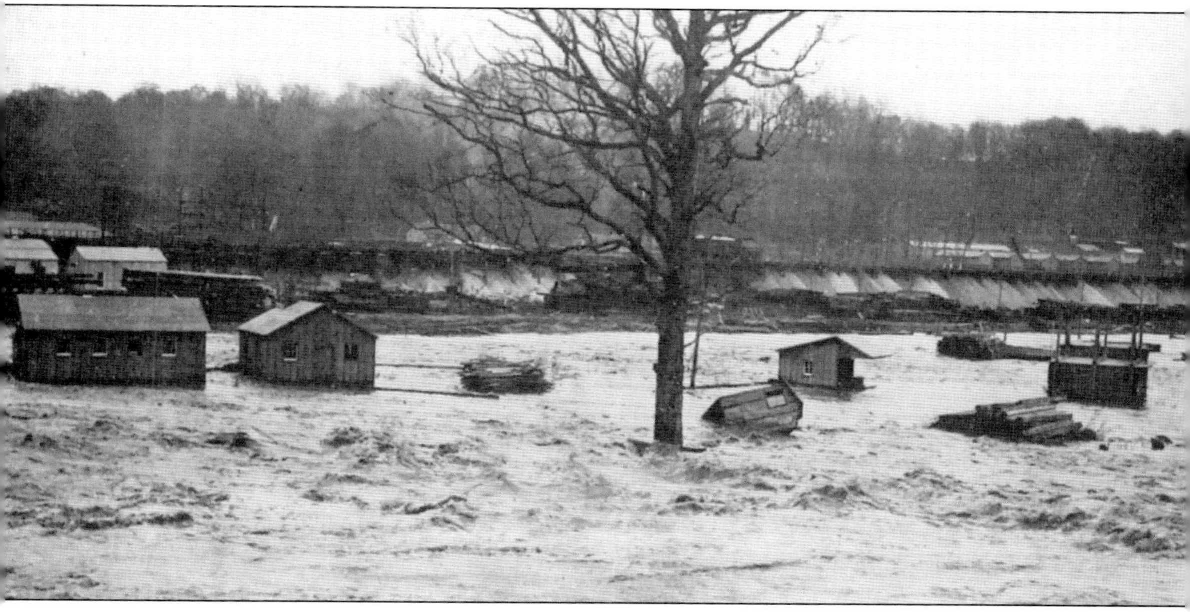

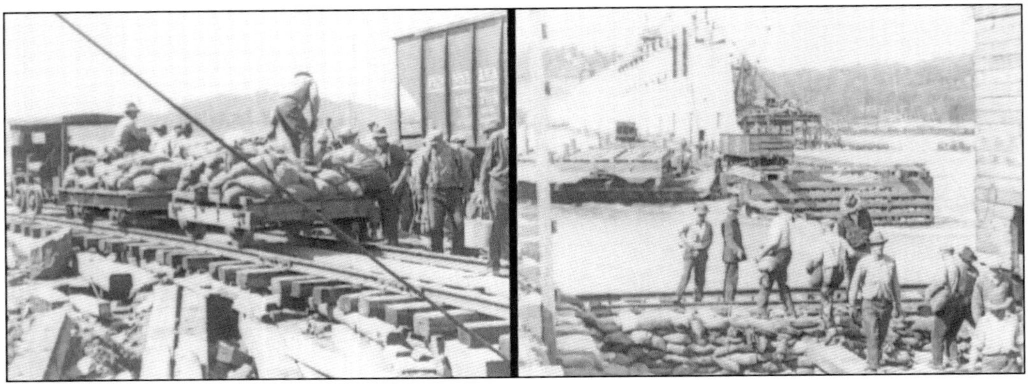

During the 1926 flood, the builders of the dam were most worried that the powerhouse cofferdam would be inundated, for if the cofferdam flooded, much valuable time would be lost. Sandbagging crews worked to construct a protective wall against the flood, but the river continued to rise and eventually overtopped all barriers except the powerhouse cofferdam, where water rose to within three inches of the top before receding. The sandbaggers won the battle, but the cofferdam suffered 11 leaks that, according to newspaper accounts, would "delay indefinitely work upon the largest hydroelectric project in the East."

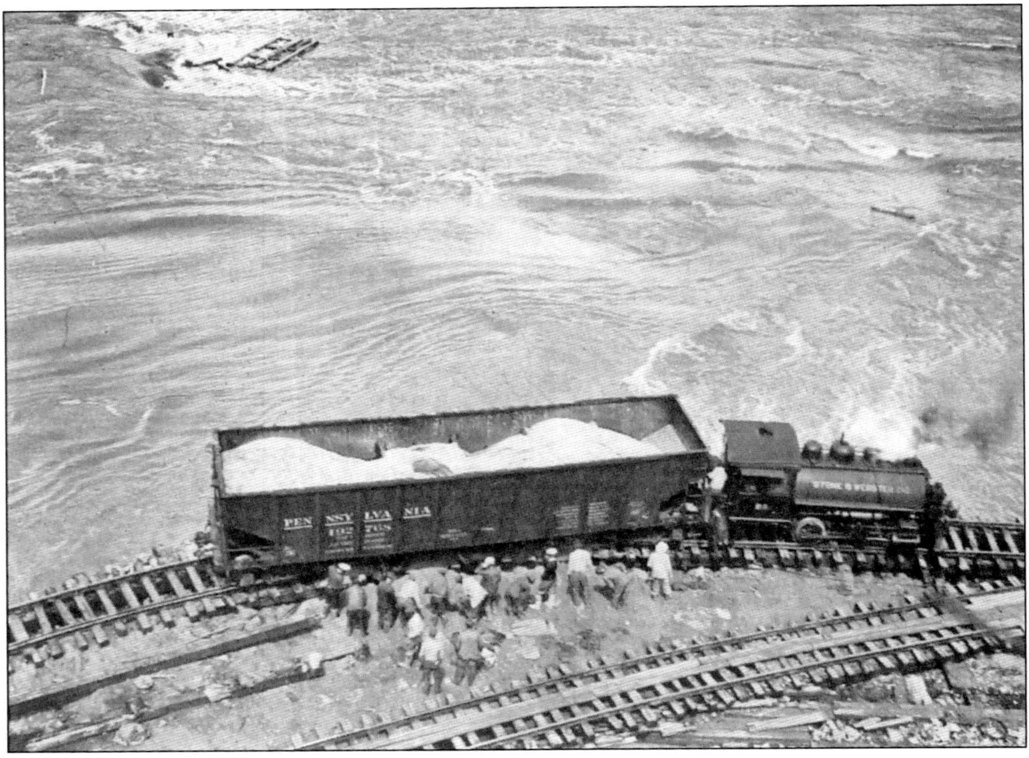

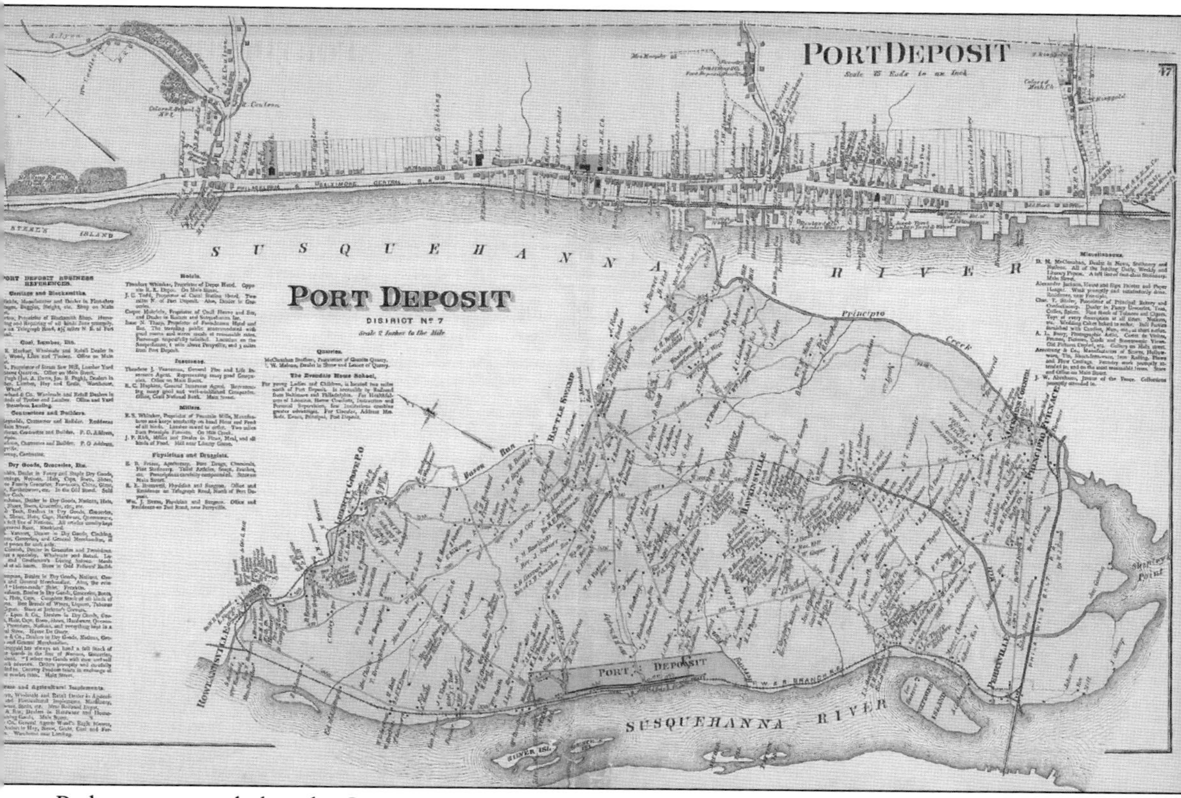

Perhaps no town below the Conowingo Dam was more vulnerable to the recurring havoc wreaked by the Susquehanna than Port Deposit, Maryland. This 1877 map shows the town of Port Deposit hugging the Susquehanna River on the southern edge of Cecil County. Port Deposit sits at the northernmost navigable point on the Susquehanna; above it are the rapids and other hazards that thwarted Capt. John Smith's expedition in 1608. Visible on the map are the wharves that grew up to service the chief industry for which Port Deposit was named—a "port of deposit" for merchant ships sailing up the Chesapeake Bay picking up loads of lumber, grain, coal, whiskey, tobacco, and other goods. The town was a narrow ribbon of buildings pinned against the Susquehanna River by imposing granite bluffs. It also lies just a few feet above tidewater, which made it defenseless against the river's ice floes and floods. (Courtesy of the Library of Congress.)

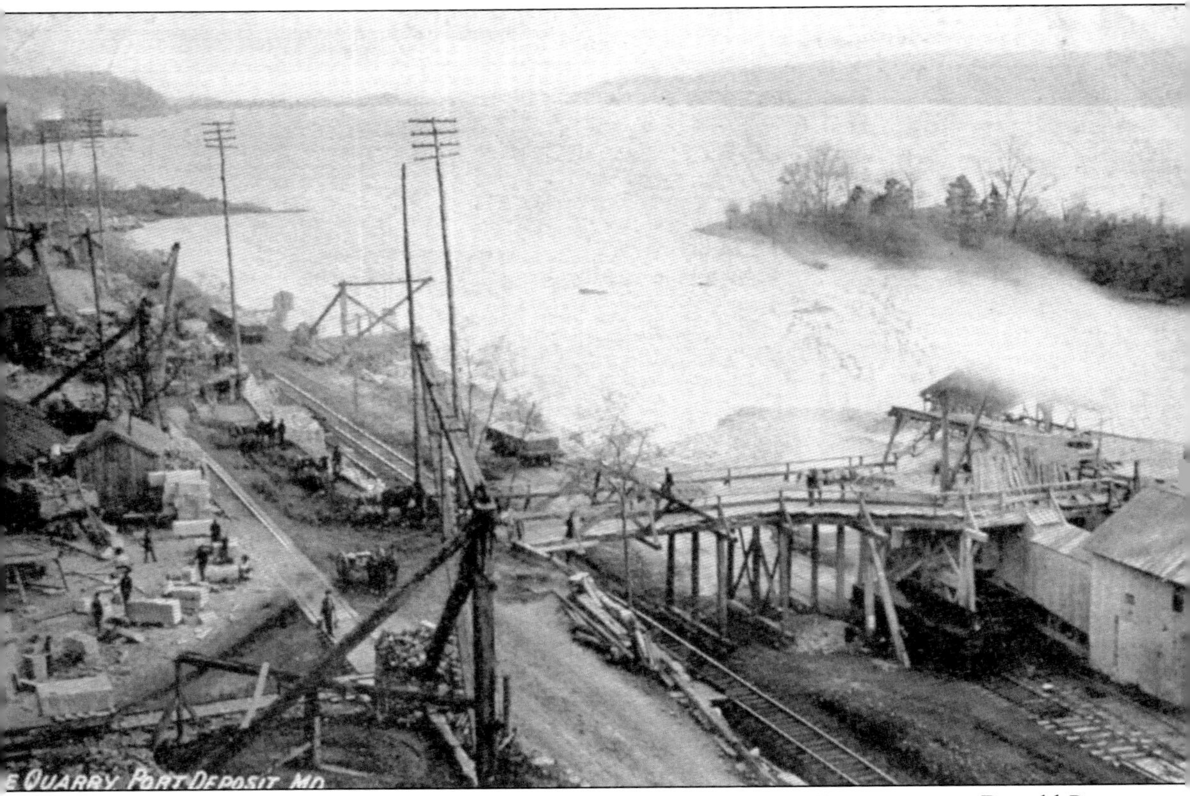

This postcard comes from the personal collection of former Port Deposit mayor Donald Poist, a descendant of Curtis Poist, who published his memories of working on the dam. The image highlights one of the chief industries of Port Deposit, the quarrying and exporting of granite. This view from a nearby bluff shows the Columbia & Port Deposit Railroad as it cut through town, past the quarry operation where large blocks of granite were loaded onto railcars and shipped out. The photograph also illustrates the proximity of Port Deposit to the river. The town relied on the slightly elevated railroad berm to act as a levee to hold the water back, but it routinely overflowed the tracks and inundated the town. (Courtesy of Donald Poist.)

This 1918 report describes fearful residents "driven from their homes" by flooding and a "huge nine- or ten-foot long ice gorge expected at any moment." An earlier report from 1808 noted that the town of Port Deposit was partially submerged in ice, water, and mud for over a month due to ice gorges. The 1918 flood caused the big warehouse of the Tolchester Steamship Company to go down in a spectacular fiery blaze. (Courtesy of the *Baltimore Sun*.)

PORT DEPOSIT FLOODED; ICE GORGE ON THE WAY

Hundreds Driven From Homes In Early Morning—Conditions Worse Than For Years And Residents Are Fearful.

Port Deposit, Md., Feb. 17.—Port Deposit was flooded, and the entire population of the northern and southern ends of the town were driven from their homes shortly after 1 o'clock this morning when the Susquehanna overflowed its banks and, by 2 o'clock had filled the streets of the town with three to five feet of water.

At that time word came to the town that the water was pouring six and eight feet over the McCall's Ferry dam. The huge nine or ten-foot ice gorge was expected any moment, and the residents of the city gathered in the center and higher part of the town, were fearful of the consequences. The conditions, it was said, are worse than for many years.

Some hint of what was coming was given yesterday afternoon, and a few of the wiser residents in the lower sections removed their furniture then. The great majority of people, however, took a chance, and suffered losses this morning.

As the flood was rising, the big warehouse of the Tolchester Steamship Company caught fire and was destroyed by a spectacular blaze.

About six miles above McCall's Ferry the ice has formed a gorge eight or nine feet high, and is still coming down stream. Although the water is not rising in any great degree it is thought that, if this continues, there is a possibility that when the gorge finally breaks at Safe Harbor the McCall's Ferry dam will be in danger of breaking away. William V. Spencer, at Lapidum, near the Tidewater canal, says the ice has piled up there in a gorge 10 feet high, and that some of the floating cakes measured were 25 inches thick. Mr. Spencer and other citizens along the water front say that at present they believe the ice will pass out into the bay without doing any damage.

Late yesterday afternoon traffic on the Columbia and Port Deposit Railroad was stopped for a time and no trains were allowed to leave Columbia until the water lowered.

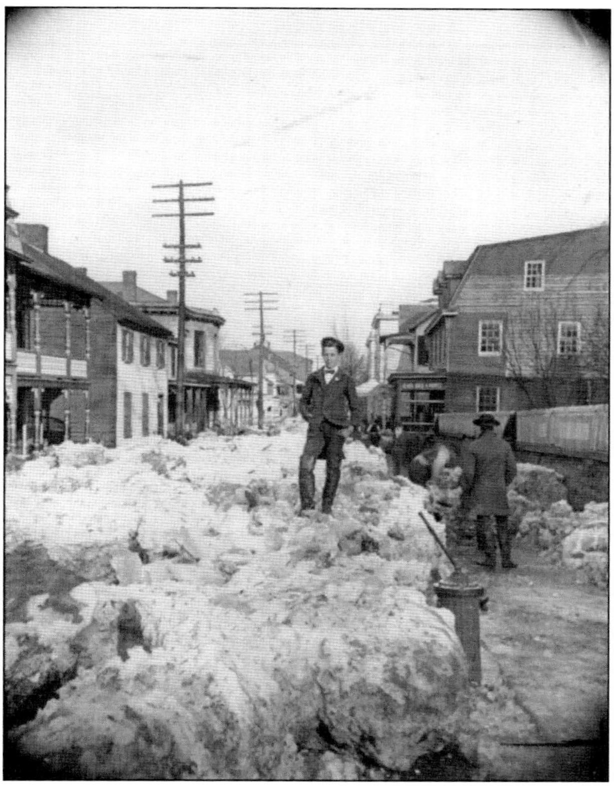

Perhaps the most famous of these Port Deposit disasters was the great ice gorge of 1904, when ice floes breached the banks and settled right into the middle of Main Street. In this photograph, presumably taken during that famous catastrophe, a young man poses stoically atop the mound of ice that landed in downtown Port Deposit. (Courtesy of the Cecil County Historical Society.)

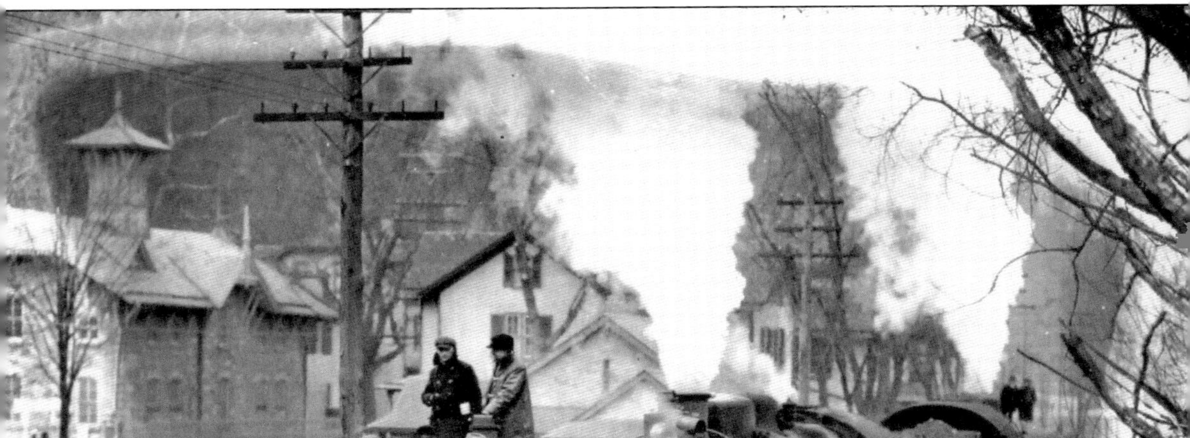

The Columbia & Port Deposit Railroad ran along the eastern shore of the Susquehanna River through downtown Port Deposit on tracks that would eventually be used to ferry supplies up to the Conowingo damsite. The elevated right-of-way formed a levee of sorts for the town, but time and again, it was breached by floods. This view of the train clearing the tracks in downtown Port Deposit after a winter flood caused by an ice gorge is thought to be from the late 1920s or early 1930s, just a few years before or after the construction of Conowingo Dam. In the background stands the town's passenger train station, a handsome granite building with decorative belfry and cornices, which had survived the devastation of ice jams for generations. (Courtesy of the University of Maryland, Baltimore County.)

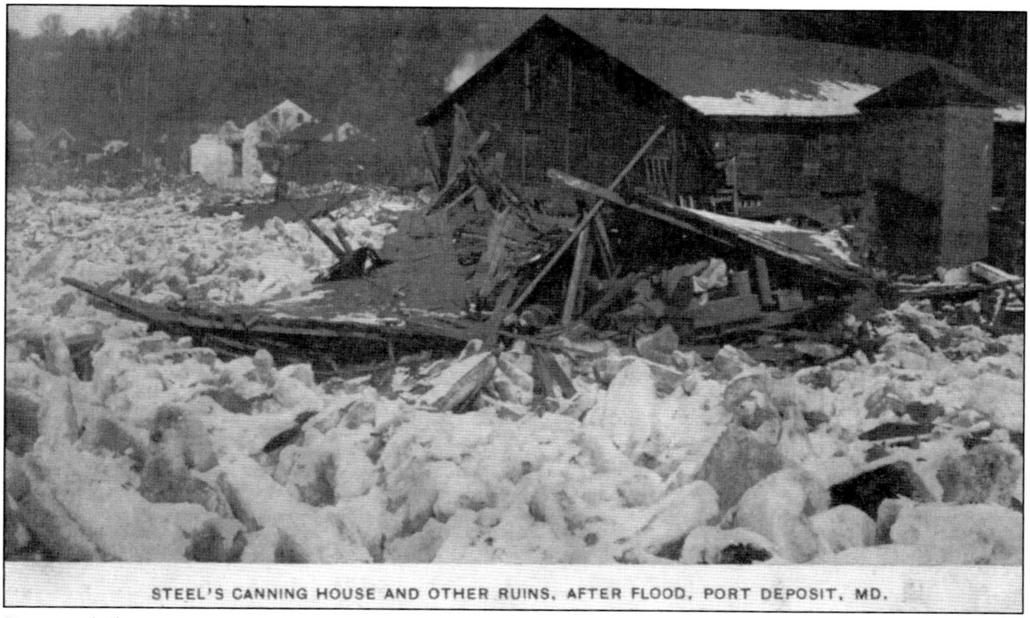

Postcards became common across the nation at the turn of the 20th century, and scenes of natural disaster were especially popular. City journalists and photographers often rushed to Port Deposit to record the spectacles that had paralyzed the town, hoping to supply their editors with headline-grabbing copy and pictures. This postcard from the early 1900s depicts the collapsed ruins of one of the town's canning houses. In 1907, *Scientific American* marveled, "It is difficult to believe that the accompanying illustrations, showing the devastation wrought by great ice masses in the Susquehanna River, are from photographs taken in Maryland, a state which is seldom associated with rigors of winter such as these. The huge masses of ice shown formed part of a great ice gorge which became jammed in the Susquehanna River, and backed up the flow of water from the freshets, causing extensive inundations." (Above, courtesy of Donald Poist; below, courtesy of Internet Archive.)

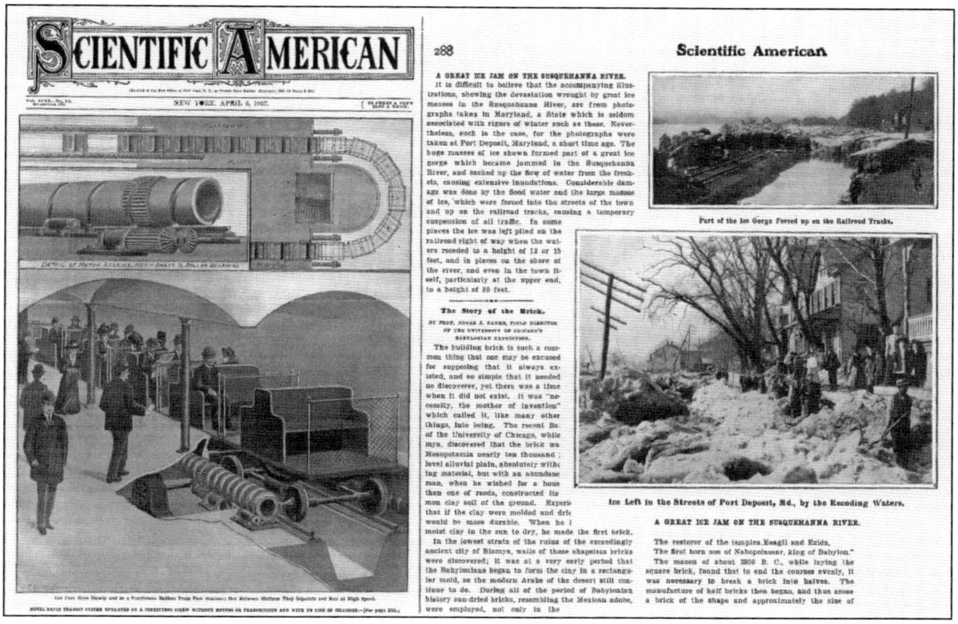
SCIENTIFIC AMERICAN

NEW YORK, APRIL 6, 1907.

288 Scientific American

A GREAT ICE JAM ON THE SUSQUEHANNA RIVER.

It is difficult to believe that the accompanying illustrations, showing the devastation wrought by great ice masses in the Susquehanna River, are from photographs taken in Maryland, a State which is seldom associated with rigors of winter such as these. Nevertheless, such is the case, for the photographs were taken at Port Deposit, Maryland, a short time ago. The huge masses of ice shown formed part of a great ice gorge which became jammed in the Susquehanna River, and backed up the flow of water from the freshets, causing extensive inundations. Considerable damage was done by the flood water and the large masses of ice, which were forced into the streets of the town and up on the railroad tracks, causing a temporary suspension of all traffic. In some places the ice was left piled on the railroad right of way when the waters receded to a height of 12 or 15 feet, and in places on the shore of the river, and even in the town itself, particularly at the upper end, to a height of 30 feet.

The Story of the Brick.

BY PROF. EDGAR J. BANKS, FIELD DIRECTOR OF THE UNIVERSITY OF CHICAGO'S BABYLONIAN EXPEDITION.

The building brick is such a common thing that one may be excused for supposing that it always existed, and so simple that it needed no discoverer, yet there was a time when it did not exist. It was "necessity, the mother of invention" which called it, like many other things, into being. The recent Ba of the University of Chicago, while mya, discovered that the brick wa Mesopotamia nearly ten thousand level alluvial plain, absolutely with(ing material, but with an abundanc man, when he wished for a hous than one of reeds, constructed its mon clay soil of the ground. Experi that if the clay were molded and drie would be more durable. When he l moist clay in the sun to dry, he made the first brick.

In the lowest strata of the ruins of the exceedingly ancient city of Bismya, walls of these shapeless bricks were discovered; it was at a very early period that the Babylonians began to form the clay in a rectangular mold, as the modern Arabs of the desert still continue to do. During all of the period of Babylonian history sun-dried bricks, resembling the Mexican adobe, were employed, not only in the

Part of the Ice Gorge Forced up on the Railroad Tracks.

Ice Left in the Streets of Port Deposit, Md., by the Receding Waters.

A GREAT ICE JAM ON THE SUSQUEHANNA RIVER.

The restorer of the temples Esagil and Ezida,
The first born son of Nabopolassar, king of Babylon."

The mason of about 2800 B. C., while laying the square brick, found that to end the courses evenly, it was necessary to break a brick into halves. The manufacture of half bricks then began, and thus arose a brick of the shape and approximately the size of

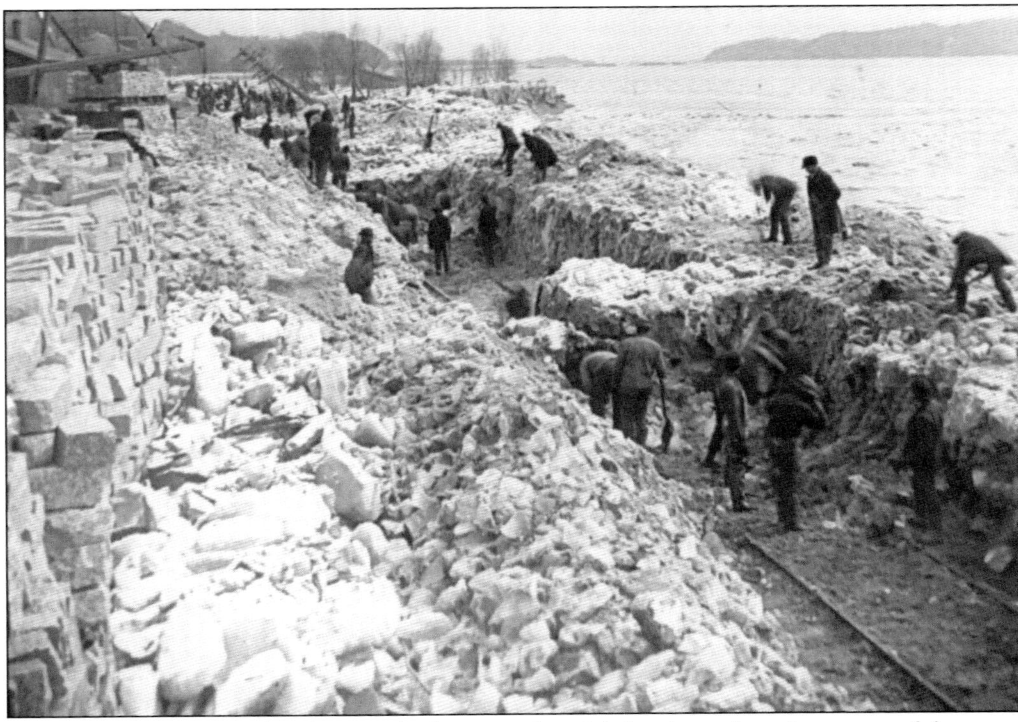

The slabs of ice were so big they crushed cars, so forceful they barged up the streets of the town, and so expansive they shoved houses off their foundations. The April 6, 1907, *Scientific American* magazine describes their stunning enormity as follows: "Ice was left piled on the railroad right-of-way to a height of 12 or 15 feet." And in the town, itself, "to a height of 30 feet," reported the magazine. The photograph above shows residents of Port Deposit clearing the train tracks beside the Susquehanna River, and below, people are posing with ice banks—nearly the height of the railcars—that had forced their way into the town. (Both, courtesy of the Cecil County Historical Society.)

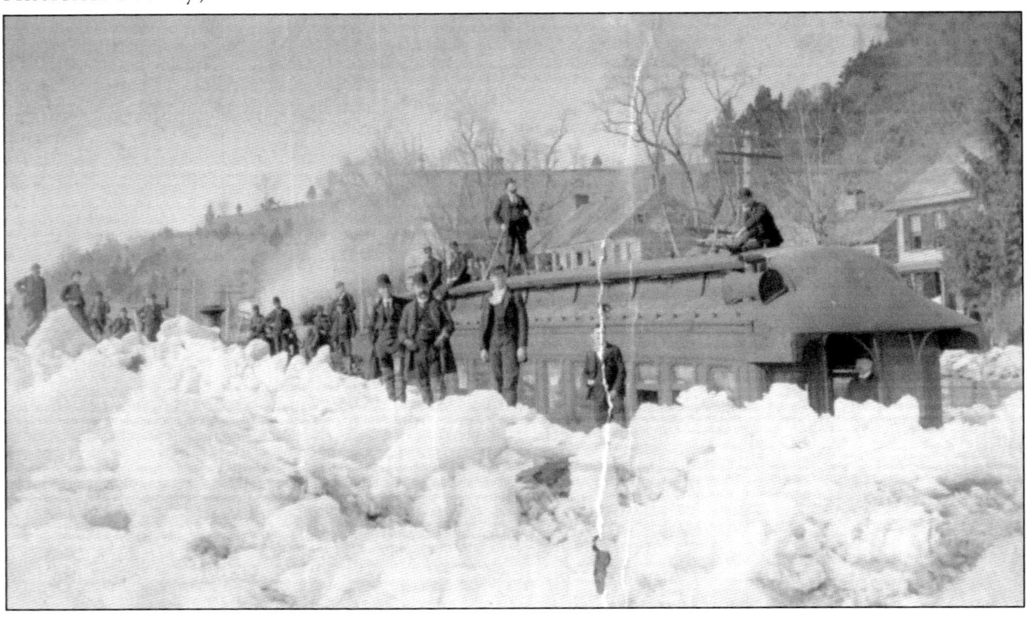

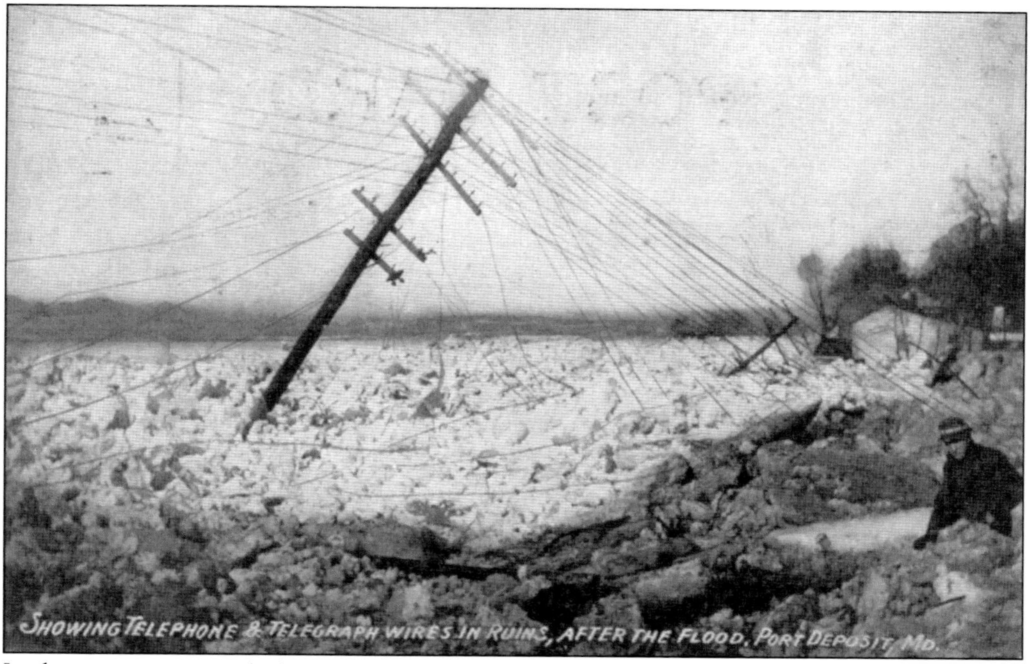

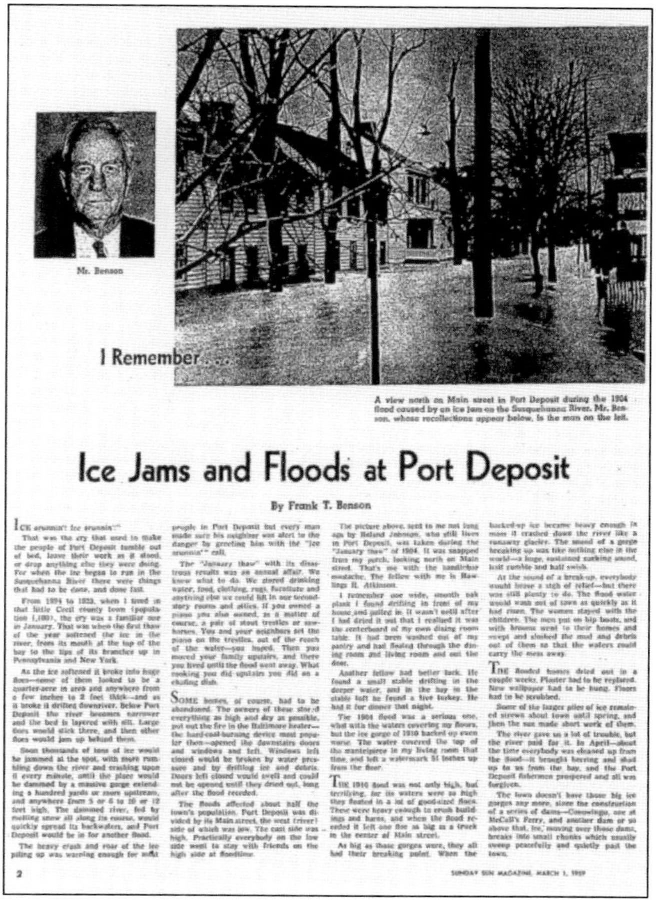

Mr. Benson

I Remember . . .

A view north on Main street in Port Deposit during the 1904 flood caused by an ice jam on the Susquehanna River. Mr. Benson, whose recollections appear below, is the man on the left.

Ice Jams and Floods at Port Deposit

By Frank T. Benson

2 SUNDAY SUN MAGAZINE, MARCH 1, 1959

In this newspaper article from 1959, Frank T. Benson, who lived in Port Deposit from 1894 to 1923, remembers the sense of dread when the frozen river began to invade the town as follows: "'Ice arunnin'! Ice arunnin'! The cry was a familiar one. For when the ice began to run in the Susquehanna River there were things that had to be done, and done fast . . . Soon thousands of tons of ice would be jammed . . . until the place would be dammed by a massive gorge . . . anywhere from five or six or 10 or 12 feet high . . . and Port Deposit would be in for another flood." Benson recalled that during the ice gorge of 1910, "the water covered the top of the mantelpiece in my living room that time, and left a watermark 51 inches up from the floor." (Above, courtesy of Donald Poist; right, courtesy of the *Baltimore Sun*.)

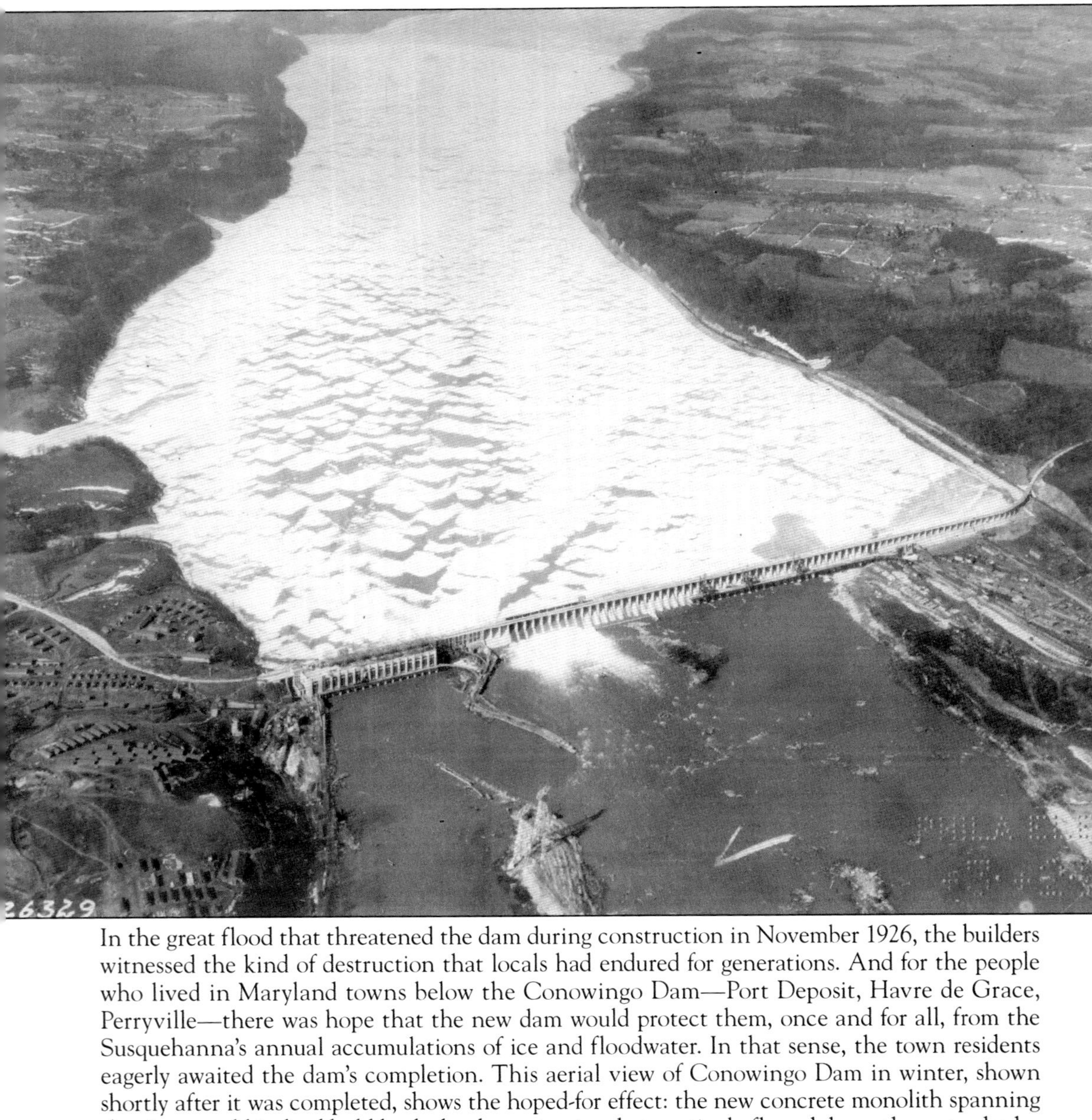

In the great flood that threatened the dam during construction in November 1926, the builders witnessed the kind of destruction that locals had endured for generations. And for the people who lived in Maryland towns below the Conowingo Dam—Port Deposit, Havre de Grace, Perryville—there was hope that the new dam would protect them, once and for all, from the Susquehanna's annual accumulations of ice and floodwater. In that sense, the town residents eagerly awaited the dam's completion. This aerial view of Conowingo Dam in winter, shown shortly after it was completed, shows the hoped-for effect: the new concrete monolith spanning the river would indeed hold back the damaging ice that routinely flowed down the watershed.

Four

THE PRICE OF PROGRESS

The Conowingo Dam created thousands of jobs for those desperately in need of them. It saved riverside towns, such as Port Deposit and Havre de Grace, from the ravages of ice floes and it provided the city of Philadelphia with much-needed electricity. But undertakings of this magnitude are seldom without cost—along with the benefits of progress, what was lost?

A company pamphlet from the period boasts that the Conowingo was built "without any loss of time in labor troubles," although construction jobs at the time were typically dangerous and there were men at the Conowingo Dam who paid dearly. Safety measures on construction jobsites were few in the 1920s, and with several thousand men working at the site and living in camps on both banks of the river, typical workers were routinely put in harm's way. There were frequent "broken bones, fractured skulls, amputations and other types of trauma," wrote local historian Mike Dixon.

The Conowingo Dam was built more than 40 years before Pres. Richard Nixon signed the Occupational Safety and Health Act into law, assuring "safe and healthful working conditions for working men and women." The act guarantees the right to a safe job, safety practices like fall protection, machine guarding, and personal protective equipment—safeguards the Conowingo Dam laborers had to live without.

Human beings were not the only casualty at Conowingo. Construction of the dam also destroyed priceless Native American petroglyphs that had been carved into the rock centuries before there was ever thought of a dam. Like the little village of Conowingo, most of these ancient messages would vanish under the rising river water when the dam was completed.

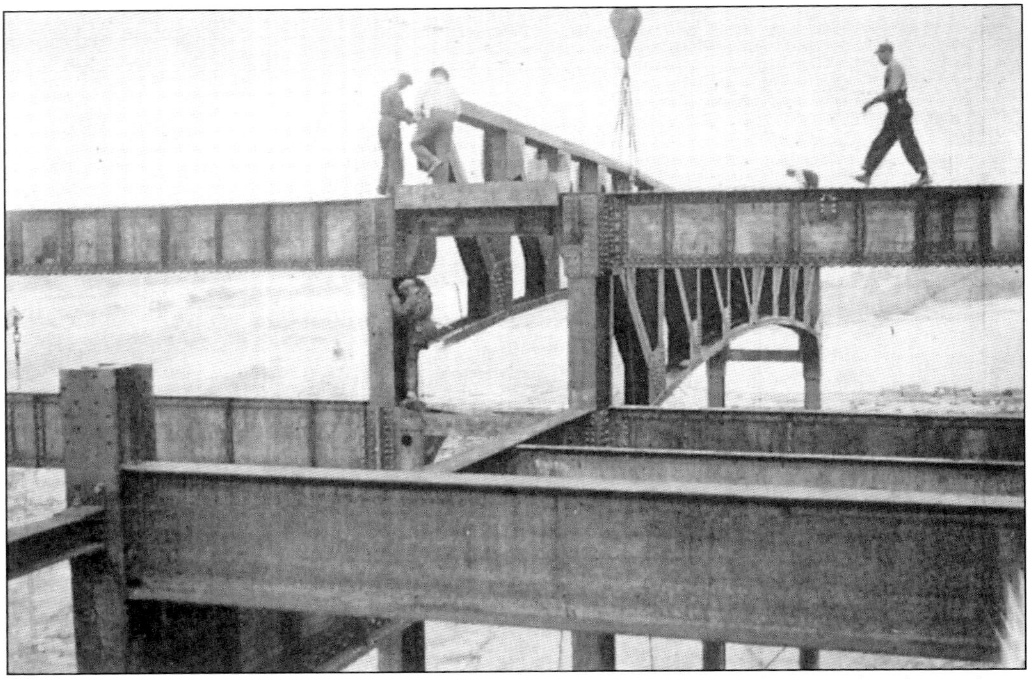

Work on the Conowingo jobsite could be life-threatening. Men labored at great heights without railings, safety restraints, or other kinds of fall protection. Construction cranes swung loads weighing thousands of pounds, steel beams, concrete chutes, and heavy timbers just above the workers' heads. Steelworkers walked atop the beams of the powerhouse seven stories up without hard hats, body harnesses, or lifelines. Injury and death could result from dynamite blasting, high-voltage electric lines, drowning, and even snakebites. There were boiler explosions, train wrecks, falls, and floods. Dixon estimates that dozens, possibly hundreds, of accident victims received on-site medical treatment in what the Philadelphia Electric Company called "a thoroughly modern hospital." But with no formal records, the toll of injuries and deaths is unknown.

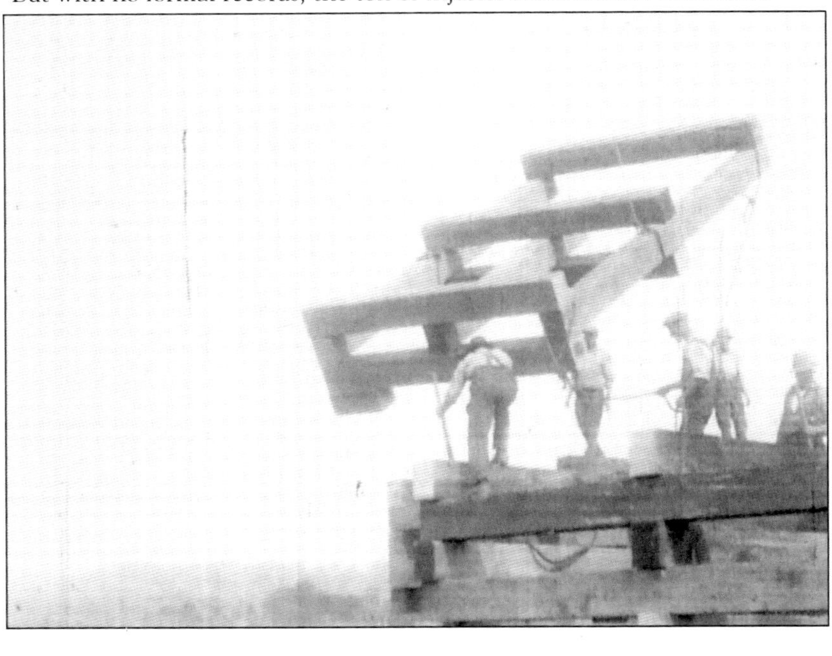

Historian Mike Dixon found evidence of 16 men killed working on the dam, and he suspects there were more. When 21-year-old Alphonso Fortier had come from Philadelphia and taken a job in Port Deposit unloading freight cars, the Conowingo project had been under way for only 12 days. He was killed after just three hours on the job when a heavy piece of machinery swerved and struck him in the chest, becoming Conowingo's first recorded casualty. (Courtesy of the *Baltimore Sun*.)

PORT DEPOSIT WORKER KILLED

Fatally Injured Three Hours After Being Employed.

Elkton, Md., March 20.—In three hours after he went to work, Alphonso Fortier, 21 years, of Philadelphia, was killed at Port Deposit, this county. Coroner Howard W. Green, of this city, gave a certificate of accidental death. Fortier had just accepted a place with contractors building the hydro-electric plant at Conowingo, and was helping to unload a derrick and other machinery from freight cars. A heavy piece of material swerved and struck him in the breast, causing an internal hemorrhage from which he died an hour later.

The coroner of Harford County, Maryland, declared the cause of death for 19-year-old Hunter Bettis was "accidental drowning while performing regular duties" at the Conowingo Dam. (Courtesy of Maryland State Archives.)

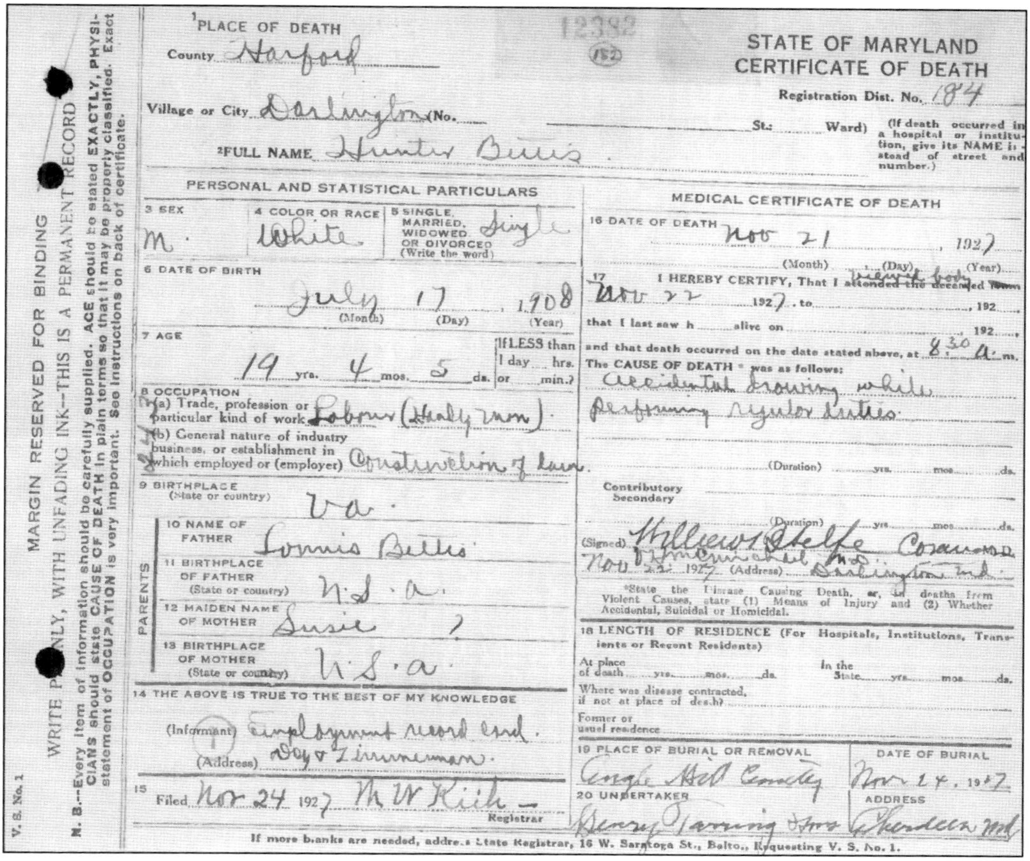

STATE OF MARYLAND
CERTIFICATE OF DEATH

1 PLACE OF DEATH
County Harford
Village or City Darlington (No. ... St.; ... Ward)
Registration Dist. No. 184
2 FULL NAME Hunter Bettis

PERSONAL AND STATISTICAL PARTICULARS
3 SEX M.
4 COLOR OR RACE White
5 SINGLE, MARRIED, WIDOWED OR DIVORCED (Write the word) Single
6 DATE OF BIRTH July 17, 1908 (Month) (Day) (Year)
7 AGE 19 yrs. 4 mos. 5 ds.
8 OCCUPATION (a) Trade, profession or particular kind of work Labour (Heavy man) (b) General nature of industry, business, or establishment in which employed (or employer) Construction of dam
9 BIRTHPLACE (State or country) Va.
10 NAME OF FATHER Lonnie Bettis
11 BIRTHPLACE OF FATHER (State or country) U.S.A.
12 MAIDEN NAME OF MOTHER Susie ?
13 BIRTHPLACE OF MOTHER (State or country) U.S.A.
14 THE ABOVE IS TRUE TO THE BEST OF MY KNOWLEDGE
(Informant) Employment record and
(Address) Day & Zimmerman
15 Filed Nov 24 1927 M W Kirk — Registrar

MEDICAL CERTIFICATE OF DEATH
16 DATE OF DEATH Nov 21, 1927 (Month) (Day) (Year)
17 I HEREBY CERTIFY, That I viewed body Nov 22, 1927, to ..., 192...
that I last saw h... alive on ..., 192...
and that death occurred on the date stated above, at 8 30 a.m.
The CAUSE OF DEATH* was as follows:
accidental drowning while performing regular duties
(Duration) ... yrs. ... mos. ... ds.
Contributory Secondary
(Duration) ... yrs. ... mos. ... ds.
(Signed) William Stelfe Coroner
Nov 22 1927 (Address) Darlington Md.
*State the Disease Causing Death, or, in deaths from Violent Causes, state (1) Means of Injury and (2) Whether Accidental, Suicidal or Homicidal.
18 LENGTH OF RESIDENCE (For Hospitals, Institutions, Transients or Recent Residents)
At place of death ... yrs. ... mos. ... ds. In the State ... yrs. ... mos. ... ds.
Where was disease contracted, if not at place of death?
Former or usual residence
19 PLACE OF BURIAL OR REMOVAL Eagle Hill Cemetery
DATE OF BURIAL Nov 24, 1927
20 UNDERTAKER Henry Tarring Bros
ADDRESS Aberdeen Md.

MARGIN RESERVED FOR BINDING
WRITE PLAINLY, WITH UNFADING INK—THIS IS A PERMANENT RECORD
N. B.—Every item of information should be carefully supplied. AGE should be stated EXACTLY. PHYSICIANS should state CAUSE OF DEATH in plain terms so that it may be properly classified. Exact statement of OCCUPATION is very important. See instructions on back of certificate.
V. S. No. 1
If more blanks are needed, address State Registrar, 16 W. Saratoga St., Balto., Requesting V. S. No. 1.

Thirty Workmen Injured When Bus Upsets On Hill

Laborers Cut And Bruised In Accident Near Conowingo.

Thirty men received cuts and bruises yesterday morning when a motor bus operated by the United Railways between Baltimore and Conowingo skidded on an icy hill near Conowingo and was upset.

The bus was filled with Baltimore workmen who were employed in construction work on the new power dam at Conowingo.

Packard Dulaney, driver of the bus, reported to the United Railways that he had intentionally run the bus into an embankment near the brow of the hill when he felt it beginning to skid backward. He knew, he said, that if the machine got beyond his control it would upset in a ravine.

Raymond S. Tompkins, assistant to C. D. Emmons, president of the United Railways, said all the glass in the bus was shattered, but that the passengers and the driver escaped serious injury.

Most of the injured passengers were taken to Conowingo by another bus which appeared shortly after the accident. They were taken to a hospital maintained by the engineering company constructing the power dam there.

Many laborers at the dam were hired by the Arundel Corporation at its employment office in Baltimore and were then bused to the construction site some 35 miles away. In January 1927, thirty such workmen were injured when the motor bus in which they were riding "skidded on an icy hill near Conowingo and was upset." (Courtesy of the *Baltimore Sun*.)

INJURED IN TRAIN CRASH

Alvan Prather, aged 25 years, of Inwood, W. Va., while firing the engine drawing loaded cars on the Stone & Webster Company's railroad, running from Havre de Grace to the site of the Conowingo dam and power house, at Shure's Landing, was badly injured, when the engine, owing to an open switch, ran upon a siding and into a string of empty cars. The jolt crushed in the rear end of the engine, catching Prather. He was taken to the Camp hospital, where his crushed right leg was amputated, and his left one was found doubly fractured. His parents were notified and arrived on Thursday last week.

Alvan Prather of West Virginia, aged 25, happened to be caught in the rear of a runaway freight train when the engine, "owing to an open switch, ran upon a siding and into a string of empty cars." Prather's crushed right leg was amputated, and his left leg was doubly fractured. (Courtesy of Mike Dixon.)

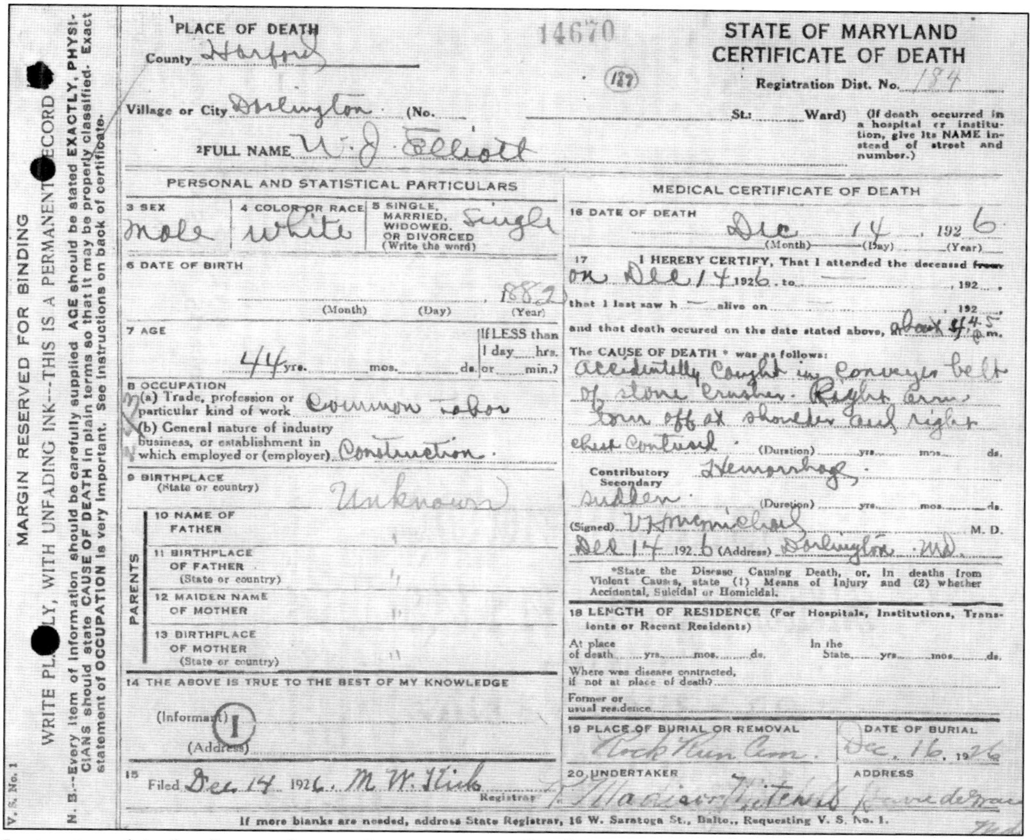

MARGIN RESERVED FOR BINDING
WRITE PLAINLY, WITH UNFADING INK--THIS IS A PERMANENT RECORD
N. B.--Every item of information should be carefully supplied. AGE should be stated EXACTLY. PHYSICIANS should state CAUSE OF DEATH in plain terms so that it may be properly classified. Exact statement of OCCUPATION is very important. See instructions on back of certificate.
V. S. No. 1

1 PLACE OF DEATH
County Harford
Village or City Darlington (No. St.; Ward) (If death occurred in a hospital or institution, give its NAME instead of street and number.)
2 FULL NAME W. J. Elliott

14670

STATE OF MARYLAND
CERTIFICATE OF DEATH
(127)
Registration Dist. No. 184

PERSONAL AND STATISTICAL PARTICULARS
3 SEX male
4 COLOR OR RACE white
5 SINGLE, MARRIED, WIDOWED, OR DIVORCED (Write the word) Single
6 DATE OF BIRTH (Month) (Day) 1882 (Year)
7 AGE 44 yrs. mos. ds. If LESS than 1 day hrs. or min.?
8 OCCUPATION (a) Trade, profession or particular kind of work Common labor
(b) General nature of industry, business, or establishment in which employed or (employer) Construction.
9 BIRTHPLACE (State or country) Unknown
PARENTS
10 NAME OF FATHER
11 BIRTHPLACE OF FATHER (State or country)
12 MAIDEN NAME OF MOTHER
13 BIRTHPLACE OF MOTHER (State or country)
14 THE ABOVE IS TRUE TO THE BEST OF MY KNOWLEDGE
(Informant)
(Address)
15 Filed Dec 14 1926. M. W. Kirk Registrar

MEDICAL CERTIFICATE OF DEATH
16 DATE OF DEATH Dec 14, 1926 (Month) (Day) (Year)
17 I HEREBY CERTIFY, That I attended the deceased from on Dec 14 1926 to 192
that I last saw h— alive on 192
and that death occurred on the date stated above, about 4.45 p.m.
The CAUSE OF DEATH* was as follows: accidentally caught in conveyor belt of stone crusher. Right arm torn off at shoulder and right chest contused. (Duration) yrs. mos. ds.
Contributory (Secondary) Hemorrhage. sudden. (Duration) yrs. mos. ds.
(Signed) V. J. [illegible] M. D.
Dec 14 1926 (Address) Darlington. Md.
*State the Disease Causing Death, or, in deaths from Violent Causes, state (1) Means of Injury and (2) whether Accidental, Suicidal or Homicidal.
18 LENGTH OF RESIDENCE (For Hospitals, Institutions, Transients or Recent Residents)
At place of death yrs. mos. ds. In the State yrs. mos. ds.
Where was disease contracted, if not at place of death?
Former or usual residence
19 PLACE OF BURIAL OR REMOVAL Rock Run Cem.
DATE OF BURIAL Dec. 16, 1926
20 UNDERTAKER [illegible] Madison Kitchell
ADDRESS Havre de Grace Md
If more blanks are needed, address State Registrar, 16 W. Saratoga St., Balto., Requesting V. S. No. 1.

In issuing this death certificate, the coroner in Darlington, Maryland, solemnly describes the gruesome accident that took the life of William J. Elliott, who was "accidentally caught in conveyor belt of stone crusher. Right arm torn off at shoulder and right chest contused." (Courtesy of Maryland State Archives.)

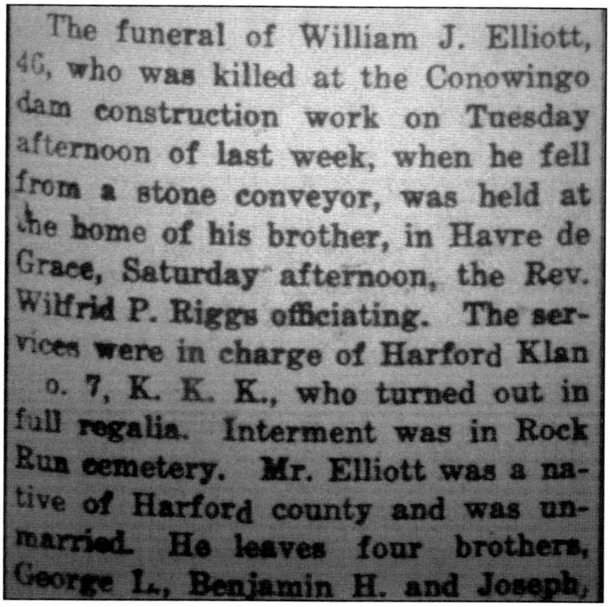

The funeral of William J. Elliott, 46, who was killed at the Conowingo dam construction work on Tuesday afternoon of last week, when he fell from a stone conveyor, was held at the home of his brother, in Havre de Grace, Saturday afternoon, the Rev. Wilfrid P. Riggs officiating. The services were in charge of Harford Klan No. 7, K. K. K., who turned out in full regalia. Interment was in Rock Run cemetery. Mr. Elliott was a native of Harford county and was unmarried. He leaves four brothers, George L., Benjamin H. and Joseph,

According to this report, the funeral services of Mr. Elliott "were in charge of Harford Klan No. 7, KKK, who turned out in full regalia." The role of Ku Klux Klansmen in the funeral is not surprising. There was a national resurgence of the organization in the 1920s, and the Klan had an especially strong presence in rural areas of Harford and Cecil Counties in Maryland. (Courtesy of Mike Dixon.)

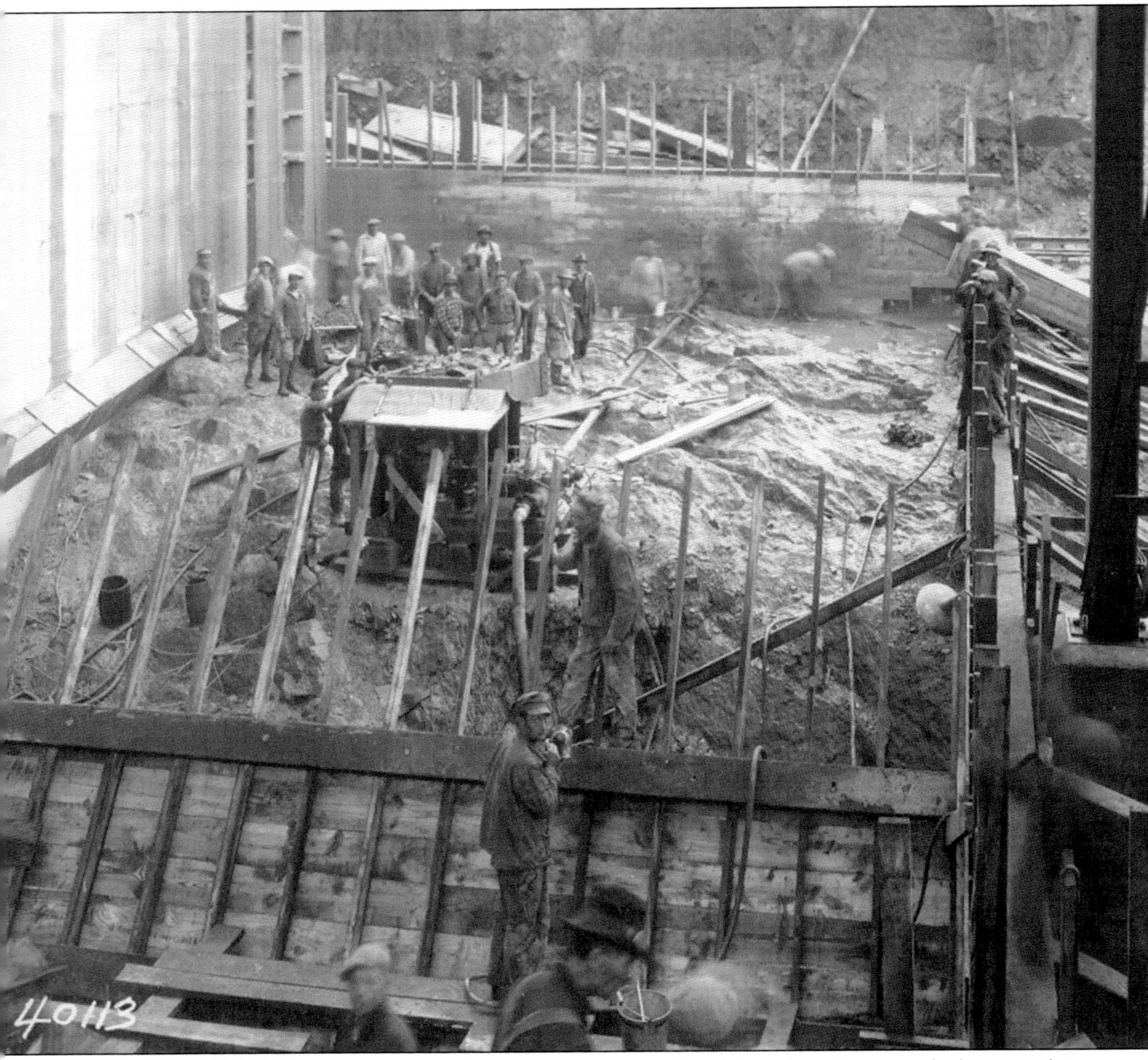

This photograph shows an integrated excavation crew with African Americans and whites working together. The Conowingo jobsite was one of the rare places in the 1920s in this part of Maryland where African Americans are known to have worked side by side with whites. However, the camps on each side of the river were white-only; blacks had to board with black families in neighboring African American communities, such as historic Berkley Crossroads and Mount Zoar, Maryland, and make the daily commute to and from the construction site. When the lunch whistle blew, meals were also segregated. The following is stated in a Philadelphia Electric Company brochure: "On each side of the river was a main mess hall with a seating capacity of 800, and on the west side a negro mess hall seating 600 was also provided."

During the early stages of work, the *Cecil Democrat* newspaper reported that "a colored laborer in stooping over a ravine to lift a heavy piece of rock was bitten on the forehead by one of those poisonous reptiles and he died in a short time." Such notices reflect the sensibility of the era—because he was African American, the deceased man went unnamed and was referred to only as "a colored laborer." (Courtesy of Mike Dixon.)

Died From Being Bitten By A Snake

Workmen engaged in clearing ground for the new dam to be built near Conowingo, are encountering trouble from copperhead snakes, which infest that section. A few days ago, a colored laborer in stooping over in a ravine to lift a heavy piece of rock, was bitten on the forehead by one of these poisonous reptiles and he died in a short time, notwithstanding medical assistance was summoned at once and everything possible was done to save the man's life.

CONOWINGO WORKERS KILLED

Havre de Grace, Md., July 18.—Stephen Collins, 28, of Baltimore, and O. P. Shelton, 32, of Florida, employed at the Conowingo dam development, were instantly killled when they fell from the crest of the dam to the rocks beneath, a distance of about 140 feet.

In this brief but harrowing notice, the *Baltimore Sun* reported that "Stephen Collins, 25, of Baltimore and O.P. Shelton, 32, of Florida, employed at the Conowingo Dam development, were instantly killed when they fell from the crest of the dam to the rocks beneath, a distance of about 140 feet." (Courtesy of the *Baltimore Sun*.)

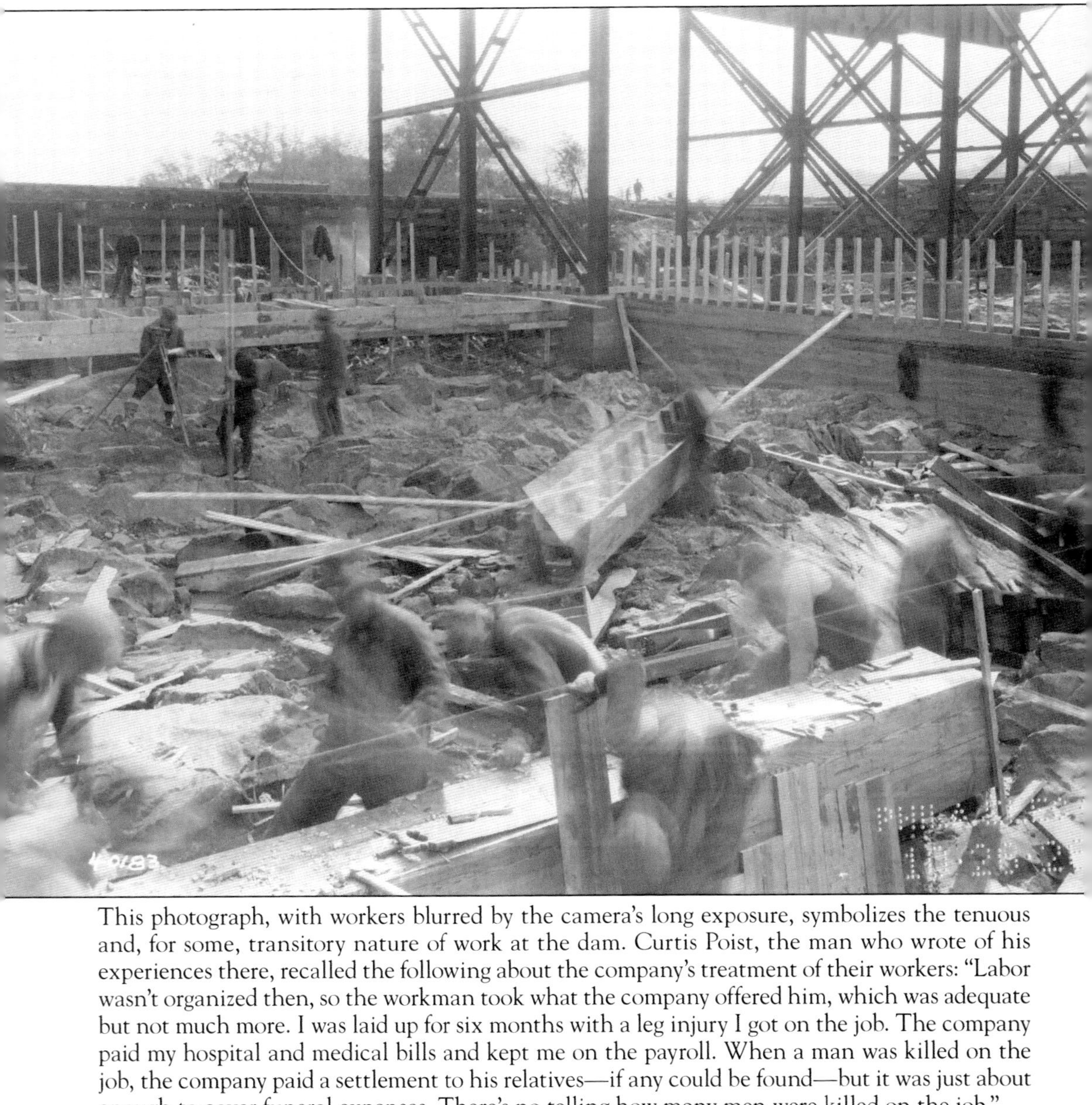

This photograph, with workers blurred by the camera's long exposure, symbolizes the tenuous and, for some, transitory nature of work at the dam. Curtis Poist, the man who wrote of his experiences there, recalled the following about the company's treatment of their workers: "Labor wasn't organized then, so the workman took what the company offered him, which was adequate but not much more. I was laid up for six months with a leg injury I got on the job. The company paid my hospital and medical bills and kept me on the payroll. When a man was killed on the job, the company paid a settlement to his relatives—if any could be found—but it was just about enough to cover funeral expenses. There's no telling how many men were killed on the job."

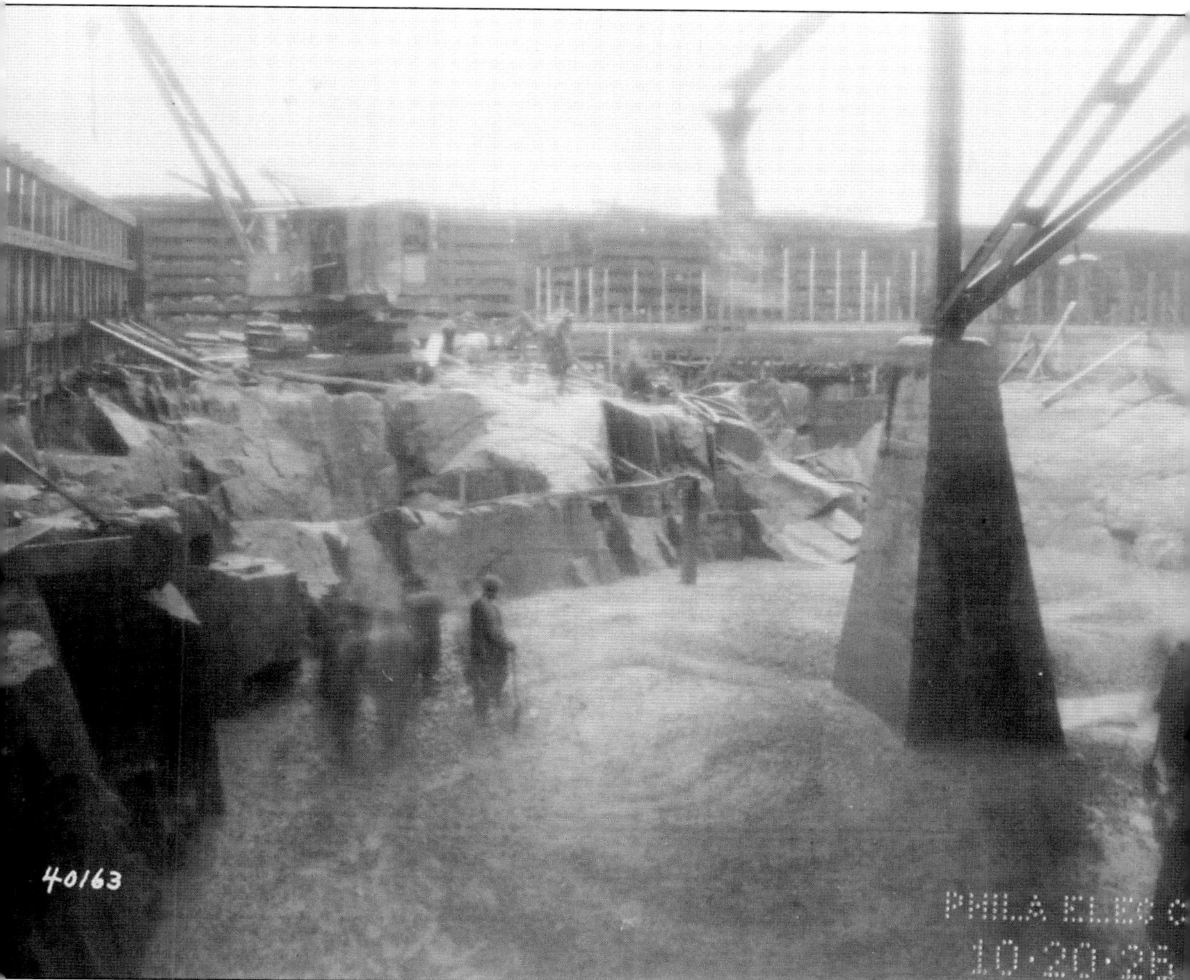

This photograph of concrete puddlers knee-deep in cement evokes an age-old legend at Conowingo, that construction workers are buried in it. As former dam workers put it, "If a worker slipped and fell in, that guy was in the concrete for good. They just kept on pouring." Conowingo was built by pouring vast sections of concrete in one continuous shot. During these pours, tremendous volumes of wet cement spewed from the chutes. "To get a worker out of there, if it was even feasible, you'd have to tear the forms apart and get rid of the concrete, and by that time, he's already been crushed." Other former workers are dubious, pointing out that the web of metal rebar reinforcing the cement would catch any worker who happened to fall in. Construction worker Curtis Poist, who was an eyewitness, tartly observed, "Usually a man was known only by the number on his badge. So, if he fell into an excavation along with several tons of wet concrete, who was to miss him, let alone mourn his passing?"

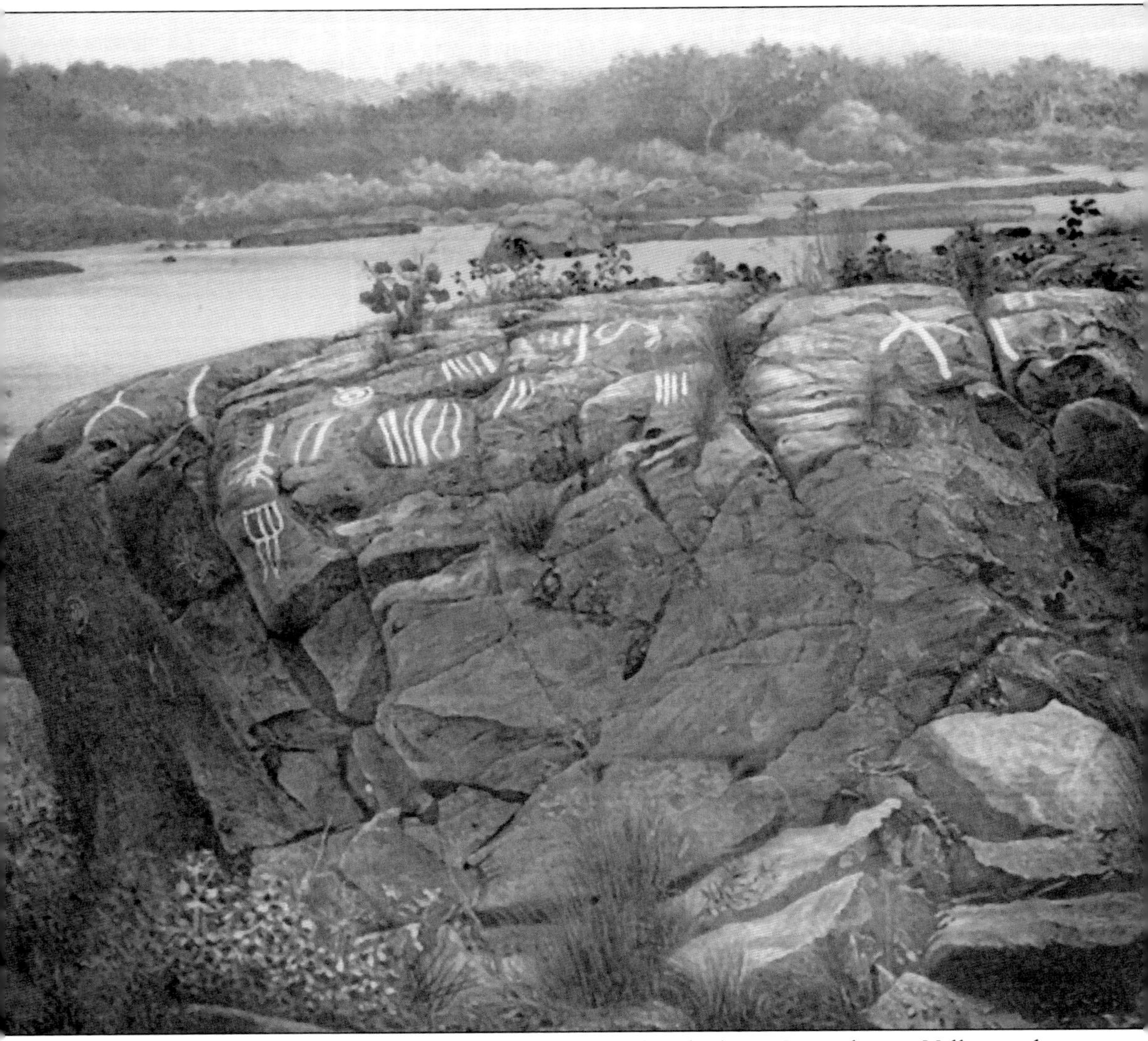

Almost 2,900 archaeological sites have been recorded in the lower Susquehanna Valley, evidence that humans were attracted to this special place long before Capt. John Smith or the Conowingo Dam. Before European contact, native Susquehannock Indians lived, worked, and fished in this hospitable location, and they left indelible evidence of their presence—petroglyphs, carved into nearby rocks and islands of the Susquehanna River. Dense concentrations of "Indian rock art" were located at Bald Friar Ford, an old river crossing just a stone's throw from the Conowingo construction site. Lying as they did behind the dam and below the eventual pond level, the Bald Friar petroglyphs would have disappeared forever if something had not been done to preserve them. (Courtesy of the Pennsylvania Historical Commission.)

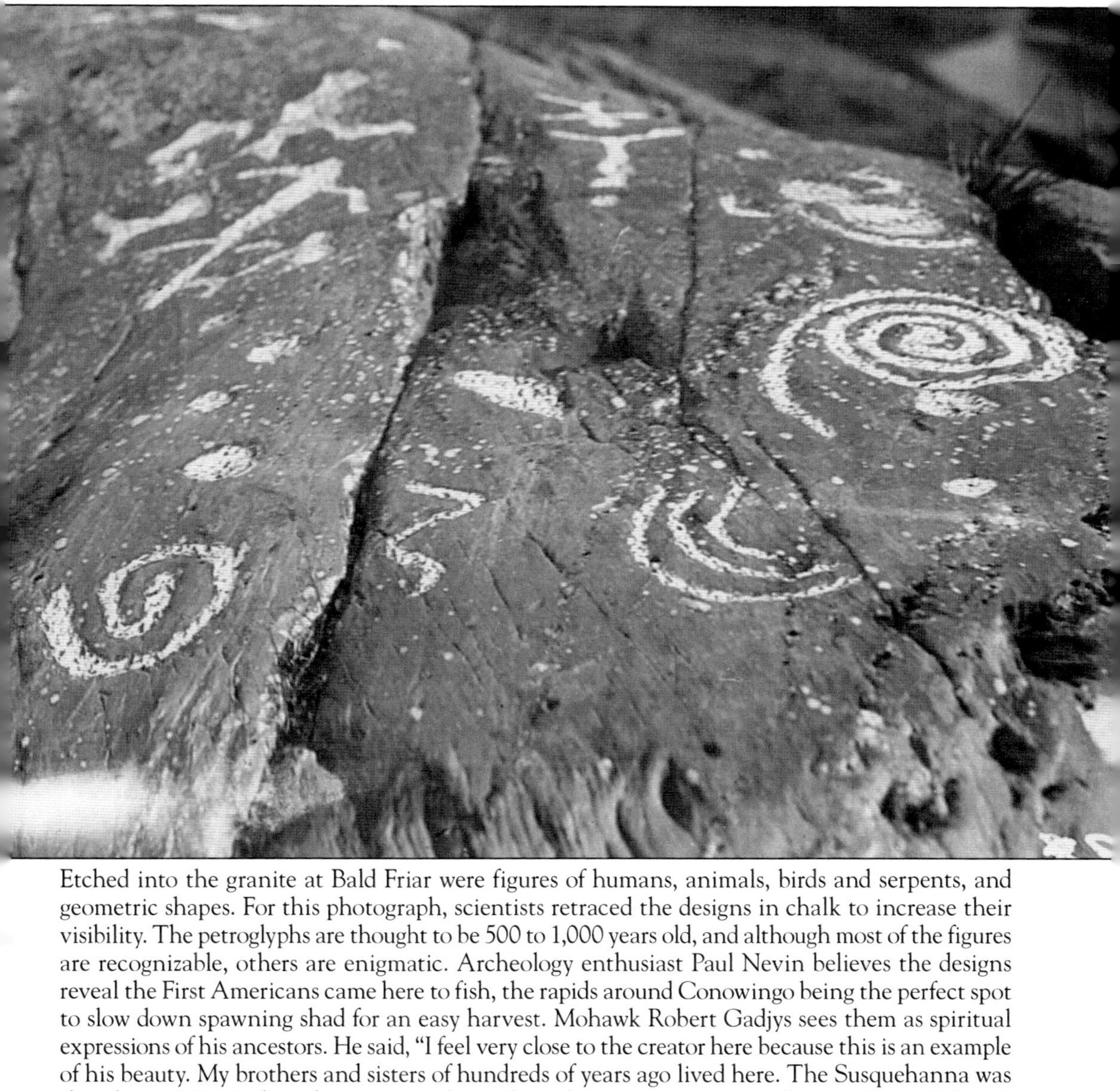

Etched into the granite at Bald Friar were figures of humans, animals, birds and serpents, and geometric shapes. For this photograph, scientists retraced the designs in chalk to increase their visibility. The petroglyphs are thought to be 500 to 1,000 years old, and although most of the figures are recognizable, others are enigmatic. Archeology enthusiast Paul Nevin believes the designs reveal the First Americans came here to fish, the rapids around Conowingo being the perfect spot to slow down spawning shad for an easy harvest. Mohawk Robert Gadjys sees them as spiritual expressions of his ancestors. He said, "I feel very close to the creator here because this is an example of his beauty. My brothers and sisters of hundreds of years ago lived here. The Susquehanna was their home." To Gadjys, the great numbers of spiral carvings are especially significant. "There's no question what that spiral means to me. The center is the grandfather, the creator, and all of us are in that circle of life. This means something to me and it also meant something to my grandfather's grandfather." (Courtesy of the Pennsylvania Historical Commission.)

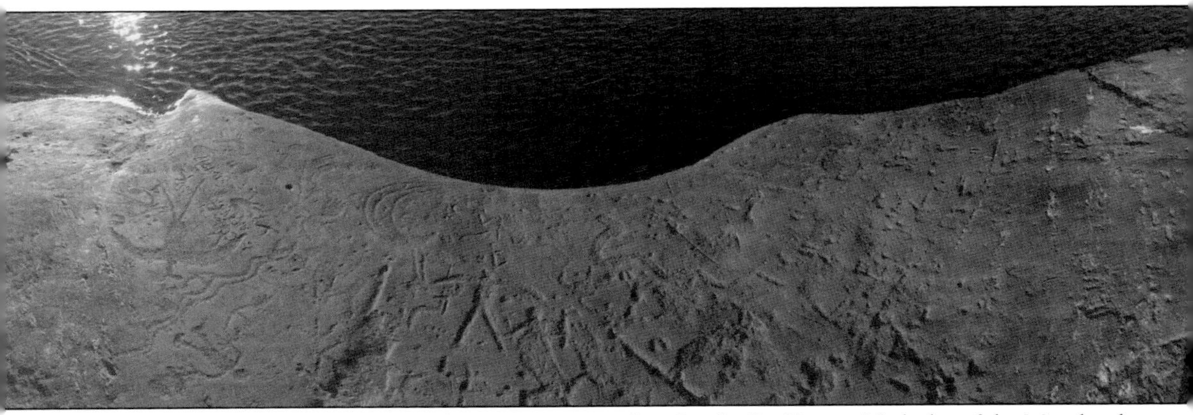

The task of saving the ancient rock carvings was undertaken by Dr. Francis Nicholas of the Maryland Academy of Sciences. Working in the shadow of the Conowingo cranes and towers, Nicholas and his team chipped and blasted into the rocks to save what they could of the petroglyphs with the hope of later reassembling the fragments like a giant rock puzzle. Dr. Nicholas, dismayed that it had come to such drastic measures, lamented his actions as follows: "It is understood, of course, that no scientist would remove the rocks unless it was absolutely necessary." But, he reasoned, "These will be lost forever unless we do something about it at once. The most important work we have now is to preserve these most remarkable of all monuments of Maryland." The team recovered about 90 pieces, a few of which now reside at Susquehanna State Park, not far from their ancestral home. (Courtesy of Mark Platts.)

Five

Dam Completion

The Conowingo Dam was built with an engineering skill and efficiency unparalleled in the history of projects of similar size. But this was the 1920s, when technological development was racing ahead and new power-generating stations were needed to fuel the race. In 1907, only 8 percent of US households had electricity; by 1930, some 68.2 percent were electrified.

The Conowingo Dam was reaching completion with equal rapidity, and in late 1927 was well ahead of schedule. Much of the main structure had risen from the Susquehanna riverbed and taken shape. Concrete pouring for the powerhouse started on December 13, 1926, and was completed five months later in May 1927. Concrete for the spillway continued throughout that year, and by January 1928, all the concrete in the dam had been laid.

Erection and assembly of Conowingo's generator room was also going full steam, and by the second half of 1927, workers were installing all the specialized mechanical and electrical components of a hydroelectric dam. In came the turbines, waterwheels, generators, floodgates, transmission towers, transformers, lightning arrestors, and oil pumps, as well as all the circuits, switches, breakers, alarms, meters, and controls necessary to make, monitor, and distribute electricity from water. A few construction statistics indicate the magnitude of the work and the immense quantities of materials. Men and machines, with the help of dynamite, excavated 348,000 cubic yards of rock and 175,000 cubic yards of earth for the foundations, and steelworkers erected 9,500 tons of structural steel in the powerhouse and embedded 6,400 tons of reinforcing steel into the concrete.

A former Conowingo Dam worker once observed, "That could never happen today." Indeed, the very idea of impounding the mighty Susquehanna by a hydro dam—and doing it so quickly—is nearly unthinkable nowadays with the requisite environmental impact studies and inevitable court battles, public protests, labor disputes, and federal inspections. The first shovel went into the ground at Conowingo in March 1926. A mere 22 months later, Conowingo Dam was making electricity.

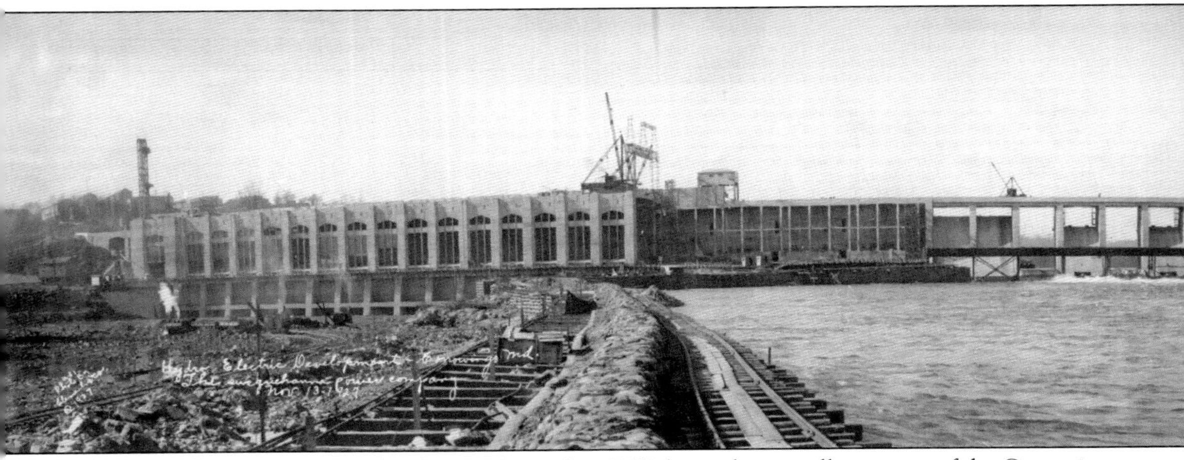

This panoramic photograph taken in November 1927 shows the overall structure of the Conowingo Dam four months before the completion date. This view from below the dam shows that, on the western shore (left), the powerhouse foundation is still being kept dry by a cofferdam, and running atop the cofferdam are temporary tracks for the railcars that continued to deliver components to the construction site. The powerhouse superstructure, made of reinforced concrete over a steel frame, is still being finished, and the building's recognizable architectural feature, the series of 16 arched window openings, are being installed. Through these windows, in the very heart of the plant, crews in late 1927 and early 1928 performed the specialized work of installing major mechanical equipment—waterwheels, shafts, turbines, and generators. On the eastern shore (right), the opening that had channeled the Susquehanna River through the middle of the dam has been

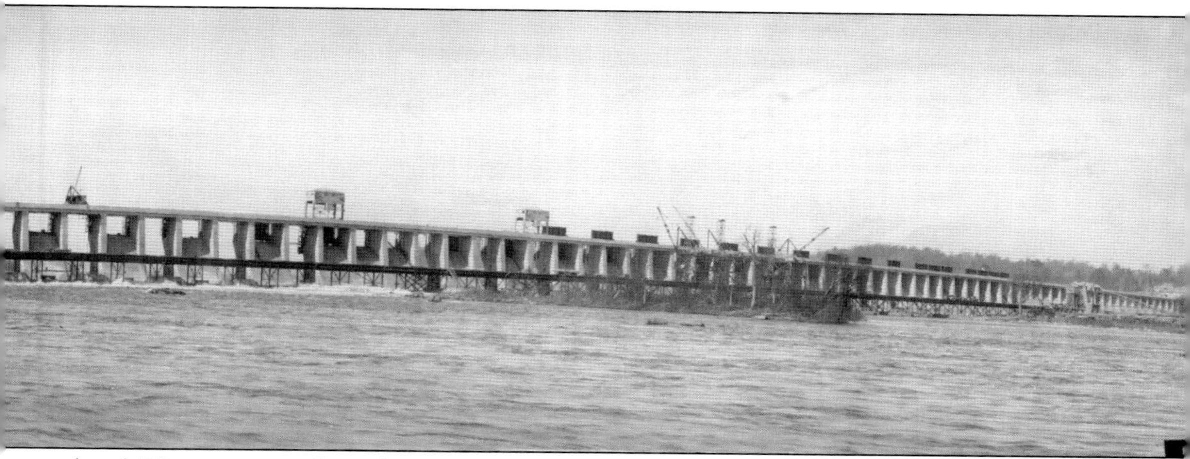

closed. The water now flows through alternating sections of the spillway that were left open for that very purpose. In a matter of months, concrete would be poured to seal these openings and permanently hold the river back. But even with these spillway sections incomplete, it was possible to complete the highway bridge on top of the dam at the same time as the rest of the project. Atop the dam, the crest gates have begun to be installed. In this photograph, the closure cofferdam that once kept the spillway section dry has been removed and water now flows freely throughout the eastern half of the site. (Courtesy of the Harford County Historical Society.)

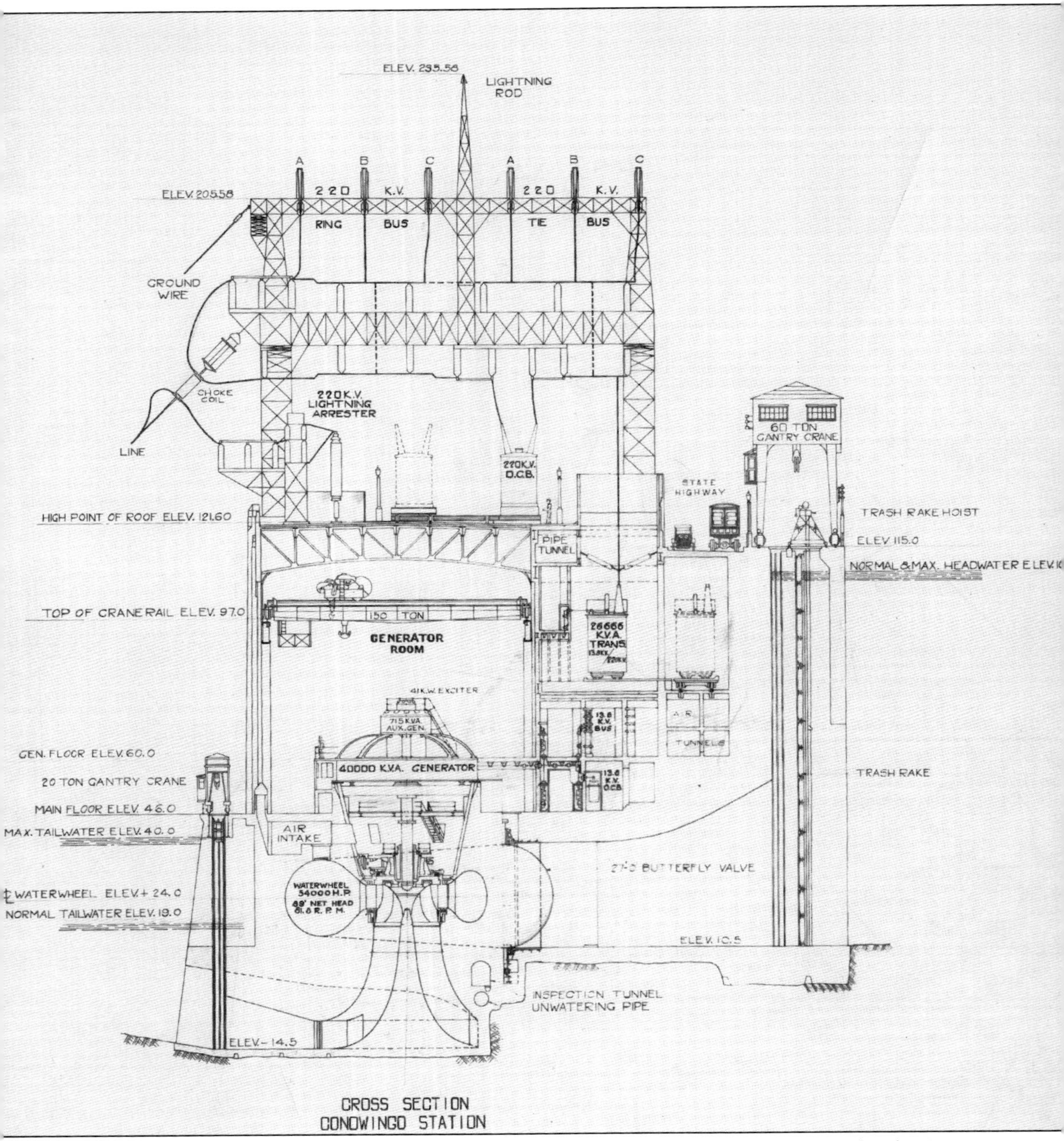

This schematic cross section of the Conowingo hydropower station illustrates the dam's power-generating components. When the giant shutoff valve opens, water rushes into the penstock from the upstream reservoir, or forebay. A scroll case is employed to equalize the pressure on the unit as water is directed to the blades of the turbine runner, or waterwheel. Water flows around the spiral scroll case, which creates a uniform force on the blades, then down through the draft tubes, and out the tailrace. This force—water dropping from a higher elevation above the dam to a lower elevation below—drives the waterwheel, which turns the shaft, which spins the generator, which produces electric current. The power is then sent to the rooftop, where transformers convert it to a higher voltage for the purpose of transmission from the plant.

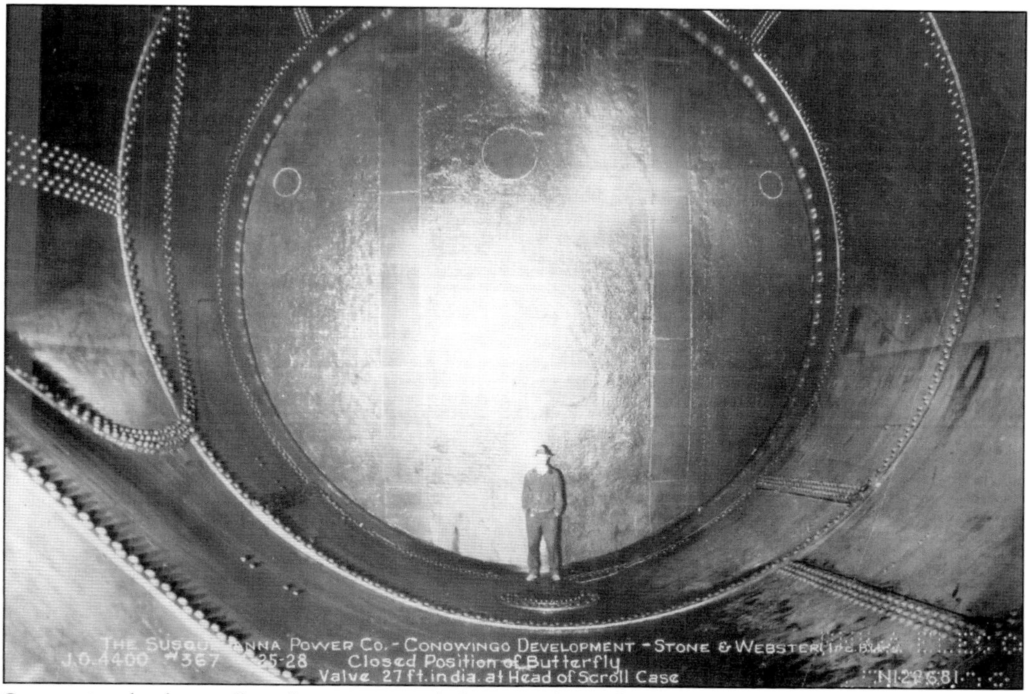

Seven circular butterfly valves prevented the river from flowing into the waterwheel when not in use or during repairs. The valves were installed at the entrance to the scroll case in each of the seven generating units and were, at the time, the largest valves ever manufactured. The giant valves, 27 feet in diameter, were fabricated in Akron, Ohio, by the B.F. Goodrich Company. Each was made of cast steel and pivoted on a vertical shaft; the disk and shaft together weighed 136 tons. They were designed to open or close in about five minutes, and when closed, they could withstand a maximum pressure of three million pounds. Goodrich engineers had developed an innovative tubular rubber gasket to seal the valve shut. Only about three inches in diameter, the gaskets were inflated like the inner tube of a tire by hydraulically pumping the tubes with water to form a nearly watertight seal.

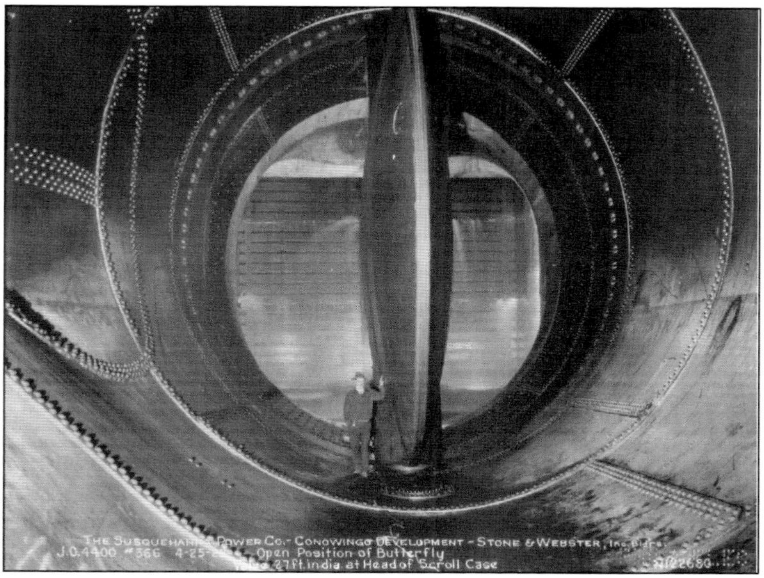

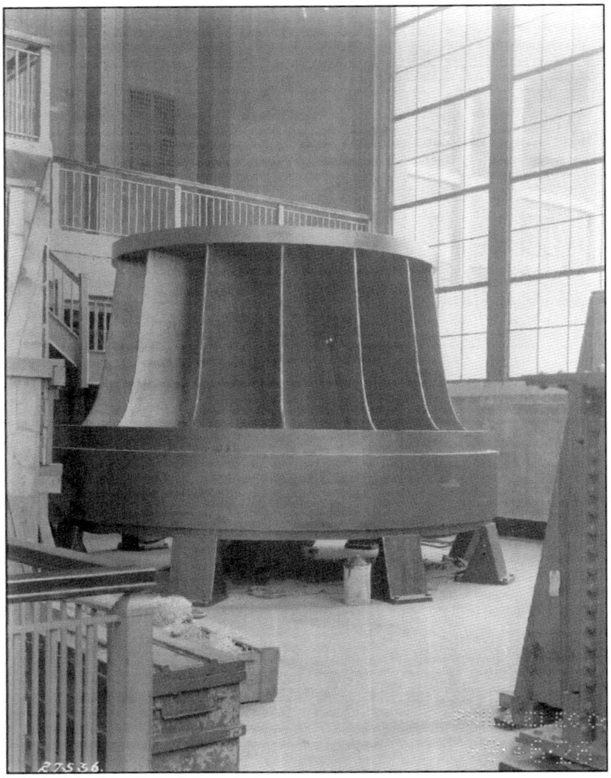

The waterwheels were the largest of their time. With a diameter at its maximum of 17 feet 9 inches and weighing about 240,000 pounds, each was rated at 54,000 horsepower,. These giant rotating components resemble a boat propeller, but instead of pushing water they are pushed by the water. Waterwheels such as the one pictured here were installed inside the scroll case and connected by a main shaft to the generator up in turbine hall.

Each shaft is made of forged steel about 36 inches in diameter—a mass of steel so thick that it would be impossible for a man to wrap his arms around it. Here, an early dam employee stands in the wheel pit near the shaft, which spins at 81.8 revolutions per minute. Just below him, the waterwheel spins inside the scroll case; above him, the generator turns inside turbine hall.

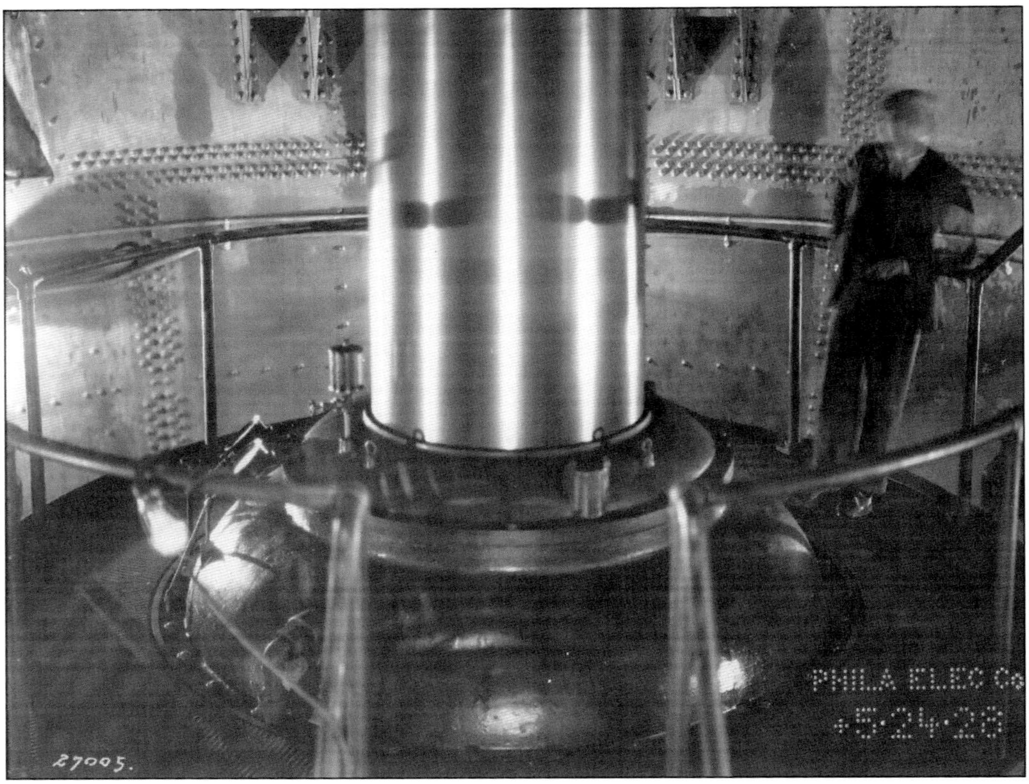

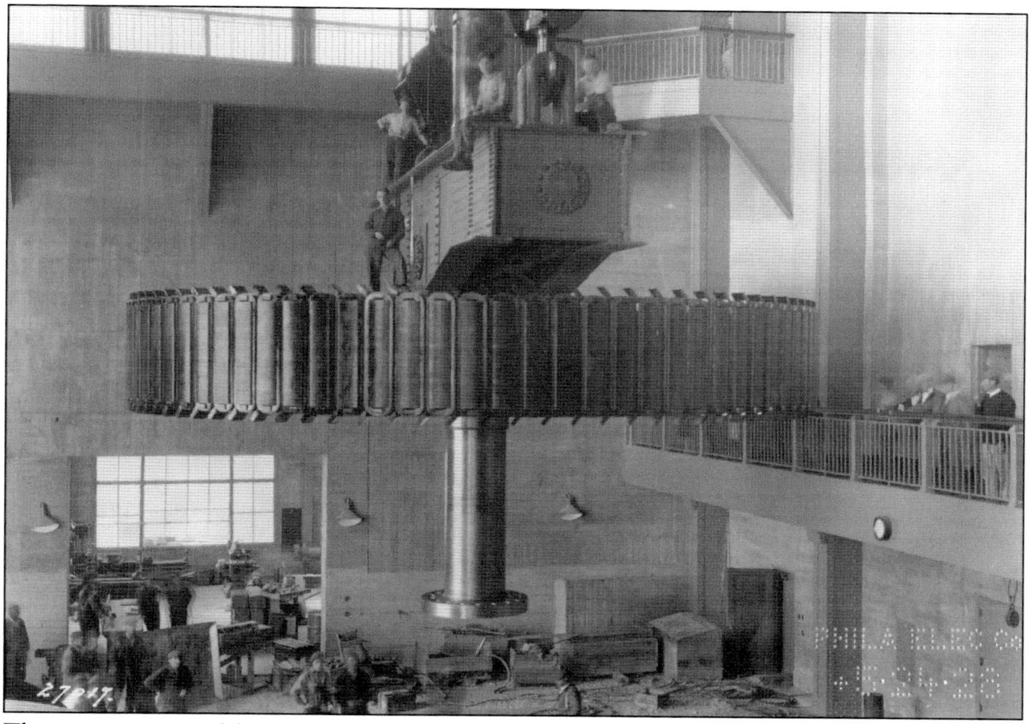

The enormous size of the generators at Conowingo are put in perspective by the photograph above, which shows intrepid workers riding atop a rotor inside turbine hall on its way to installation. The massive units were positioned by means of two traveling 150-ton cranes permanently mounted inside the generator room. To install it, the crane operator hoisted the unit, moved it out into the room, inched it over the center opening of the upper guide bearing (shown below), and lowered it down where workers secured it in place. In addition to its initial seven generators, the design of Conowingo provided for the future installation of four more units, which would provide a total capacity of approximately 600,000 horsepower. In early October 1927, workers installed these immense generators one by one into the seven guide bearings that together make up the main operating system of the Conowingo Dam.

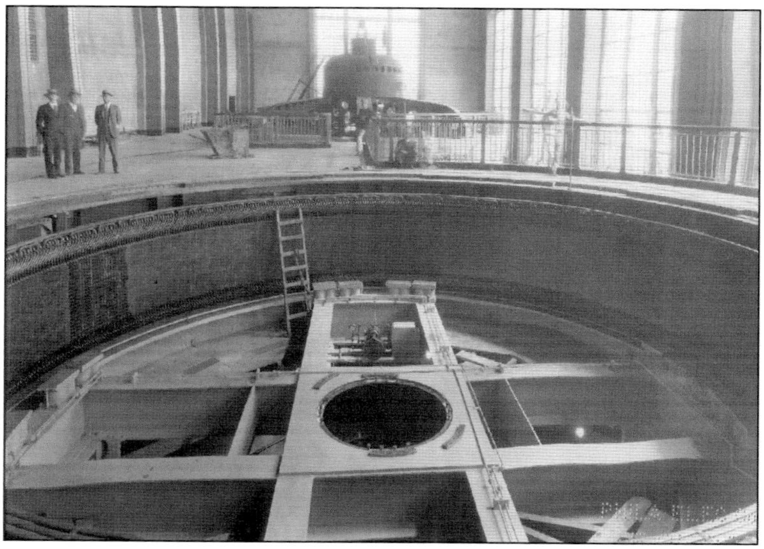

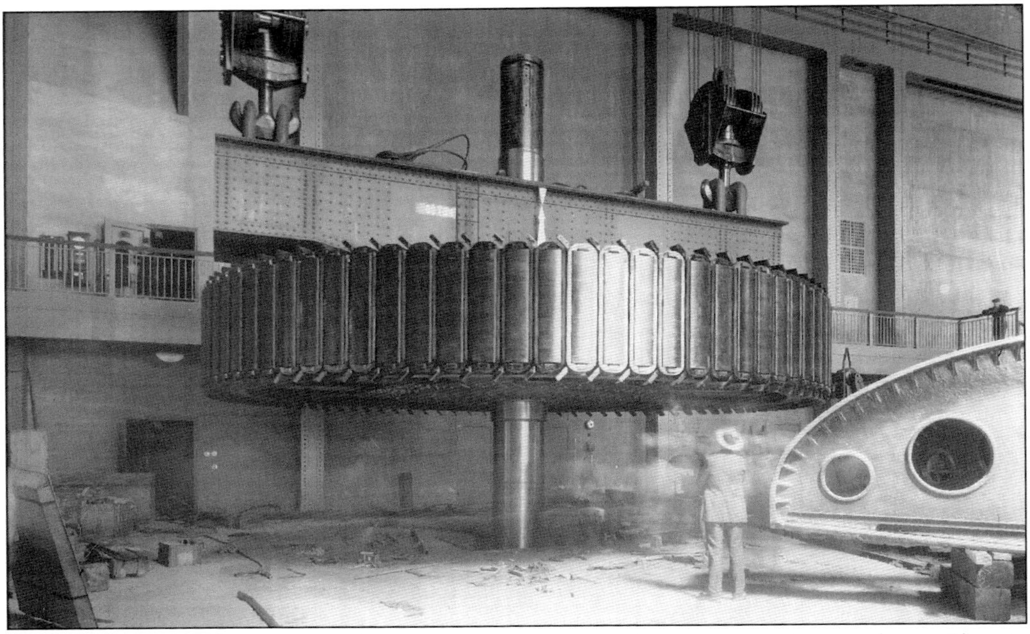

The generators were provided by both the General Electric Company and Westinghouse Electric and Manufacturing Company and, at the time, were the largest such generators ever built. Each was roughly 38 feet in diameter, 32 feet high, and weighed (together with the thrust-bearing, auxiliary unit and its exciter) well over 500 tons. The weight of the generator's thrust-bearing, including water thrust, is 1,200,000 pounds. In the photograph below, a frame spider is being craned into place. The top of the generator is mounted to this support structure, so that the generator, with its immense weight, actually hangs from above. Conowingo was outfitted with seven waterwheel units of 54,000 horsepower each for a total installed capacity of 378,000 horsepower. This was the largest amount of power, whether steam or hydro, that had ever been installed at one step in a single plant.

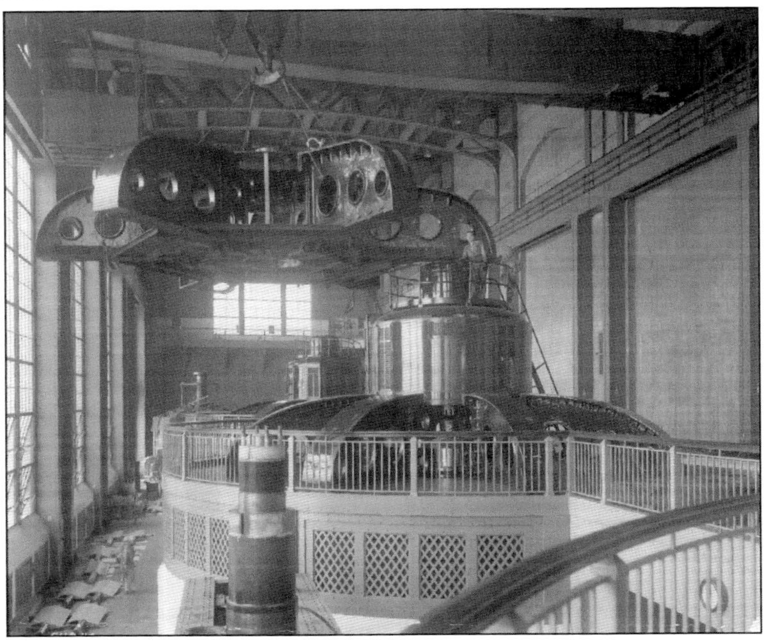

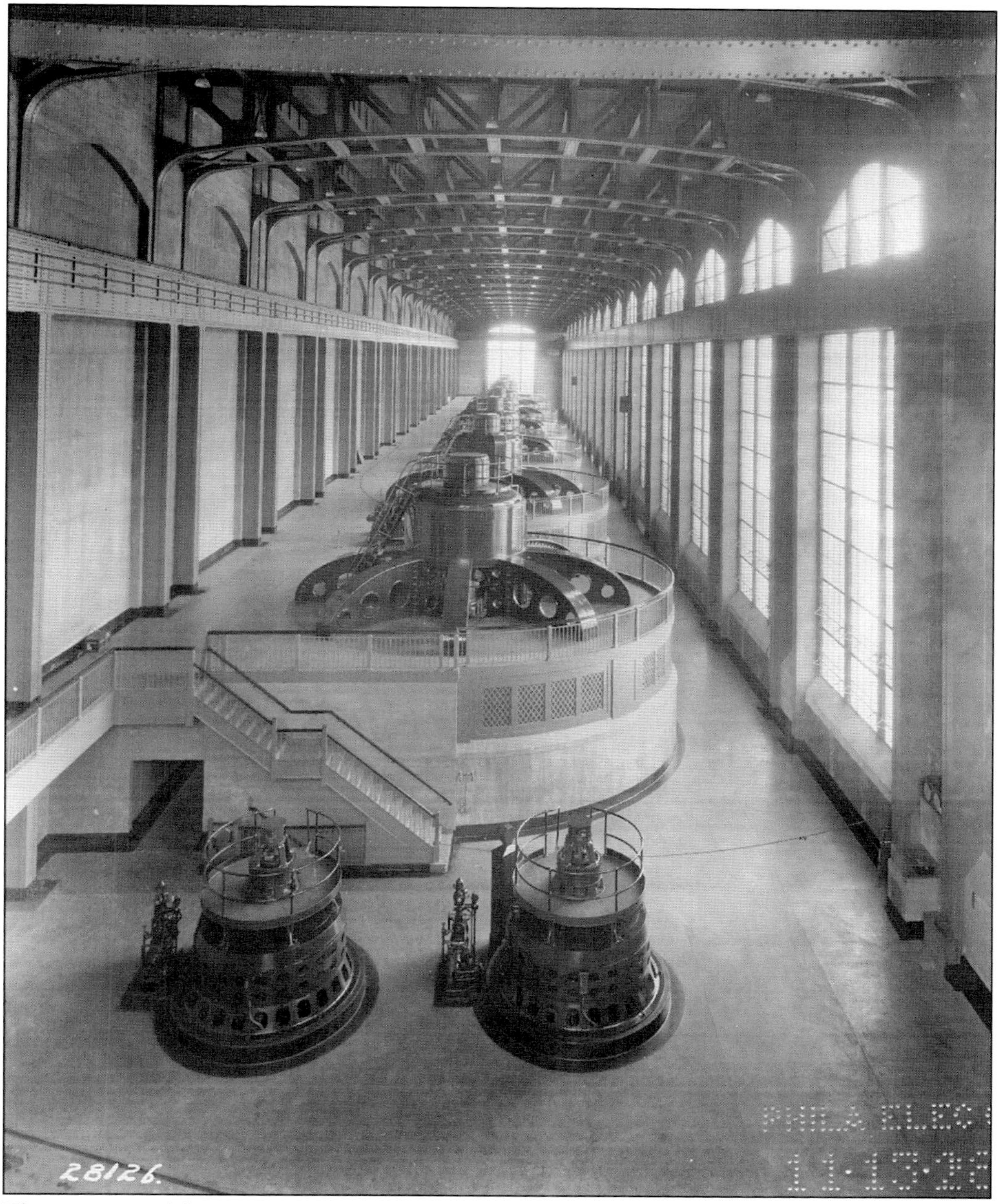

This is a view of the Conowingo Dam with its seven generator units in turbine hall and daylight streaming through the east-facing windows. Conowingo's cavernous generating room measured 70 feet wide by 75 feet high and was 620 feet long. The plant's generators can be seen receding into the distance, while in the immediate foreground are the plant's two smaller house units designed to supply operating power for the facility. Transforming the mechanical energy of the spinning turbine into electrical energy happens at the top of the generator. As the turbine spins down in the scroll case, the rotor at the top spins inside a stationary outer ring called a stator, made up of coils of copper wire called windings. Mounted on the spinning rotor are 88 poles, each made of copper windings and each causing a magnetic field. As the rotor spins inside the stator, the inner magnetic fields exert a force on the outer copper windings, resulting in pulses of electric current.

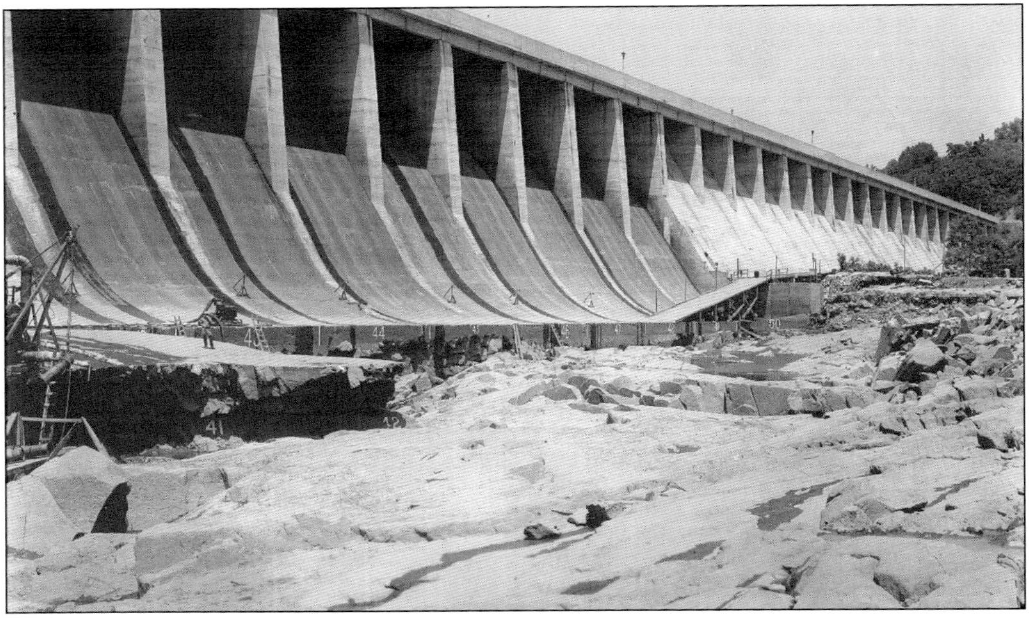

The crest of the spillway rose to a height of 86 feet (above). During ordinary times of the year, it was the spillway's function to hold back the river and channel it through the generators. However, during seasons of excess flow, when more water surged downriver than the generators could pass, one or more of Conowingo's 50 floodgates were raised, whereupon the excess water was allowed to spill through the opening and over the top of the dam. The photograph of the spillway (below), with a laborer standing below crest gate No. 45, illustrates both the thickness of the concrete at the base and how, as a concrete gravity dam, the structure sits under its own weight on the bedrock of the Susquehanna River.

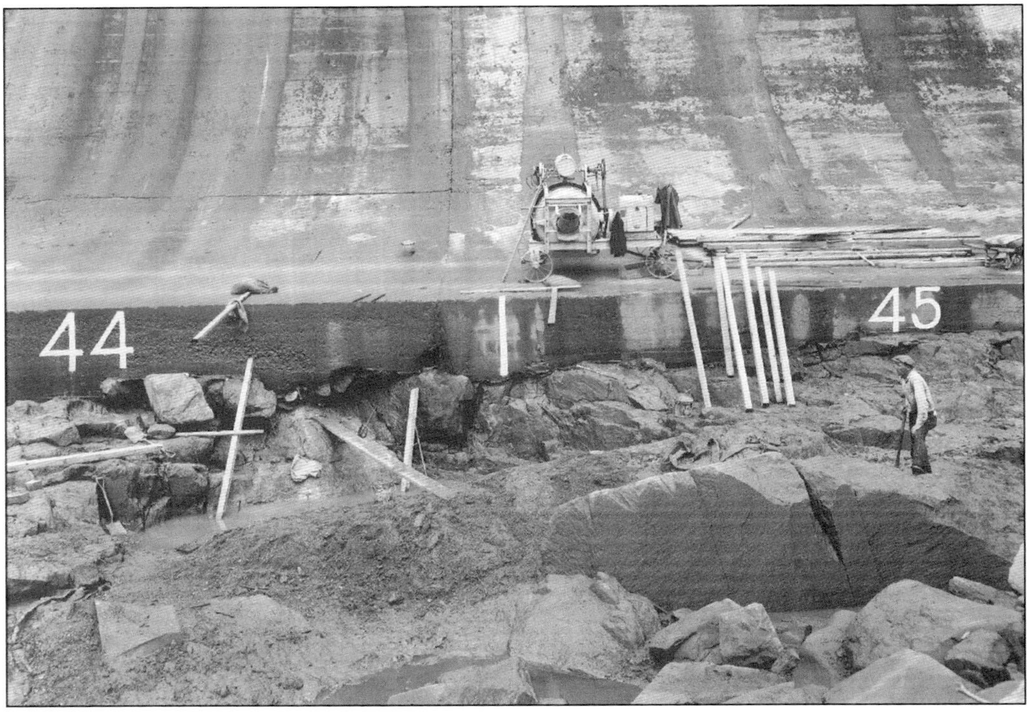

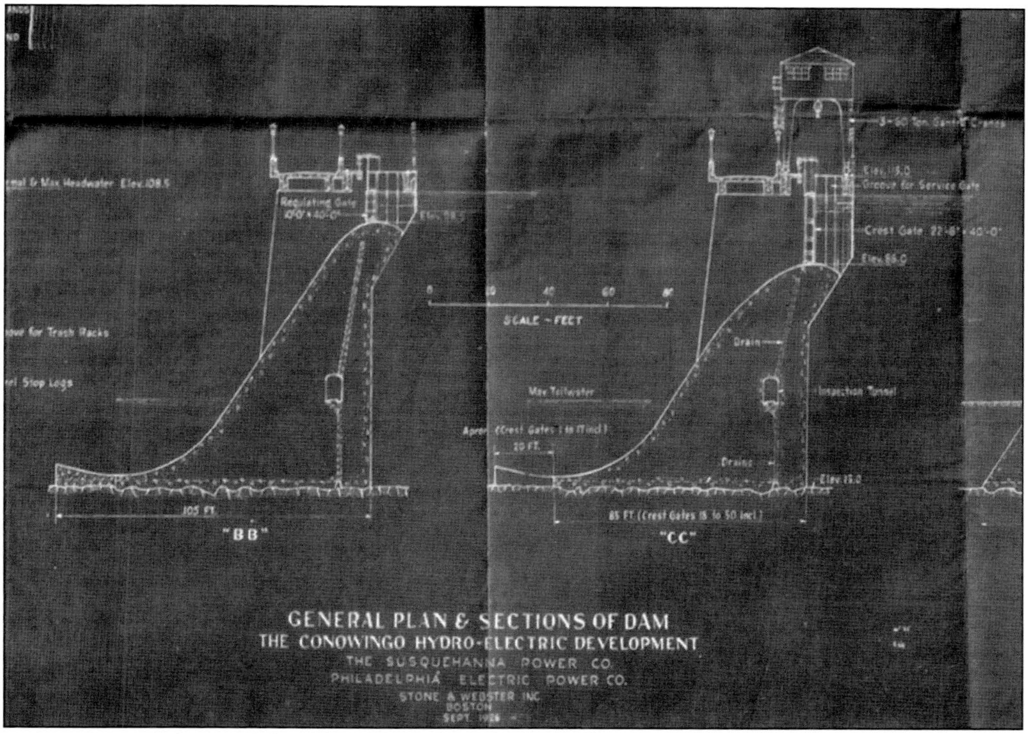

Sections BB and CC of the general plan (above) show the gentle parabola engineers designed at the toe of the spillway. As water poured down the apron of the spillway, it hit the toe with its slight 12.5-degree curve and launched upward in a powerful stream. The effect was like that of an Olympic ski jumper who hits the takeoff ramp, ascends into midair, and lands as far down the slope as possible. During the high-water season, surging water swooping off the toe of the spillway produced a visual spectacle that hydro engineers call a rooster tail (right). Tests conducted on a model of this design indicated that, with this subtle feature, maximum erosion at Conowingo would take place safely downriver some 150 feet beyond the base of the spillway.

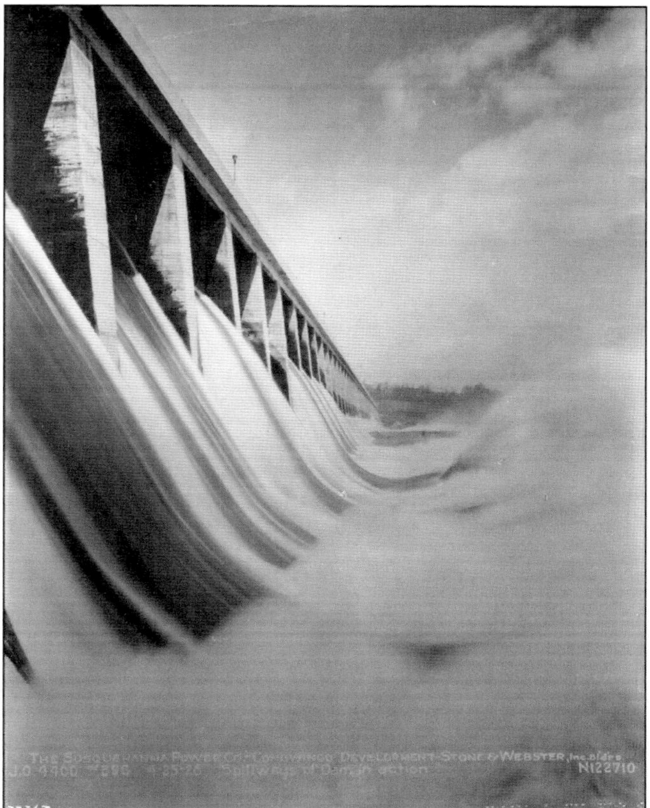

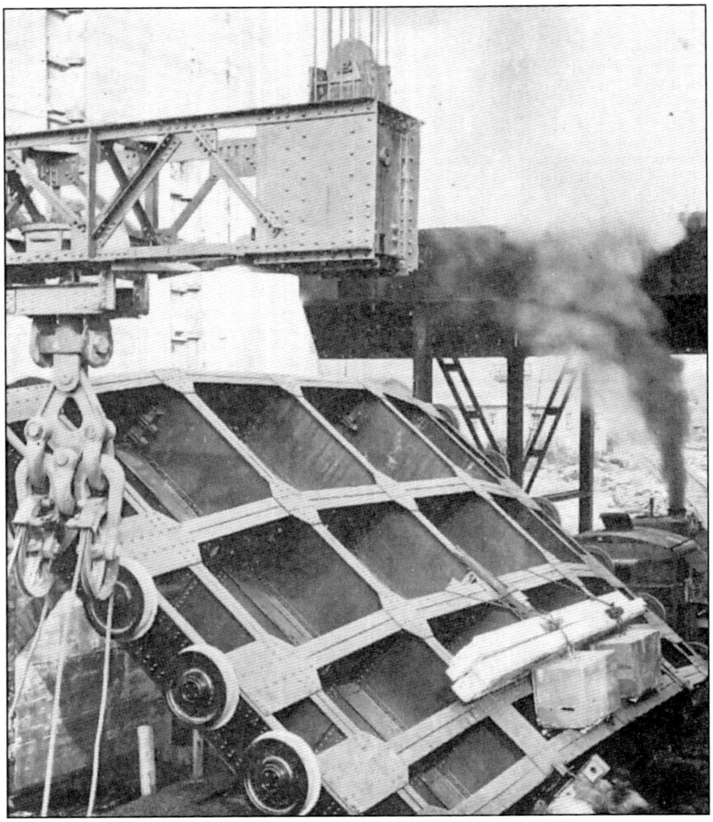

The 50 crest gates for the top of the spillway were delivered by rail up the Cecil County side of the river and hoisted by crane from the railcar into their slots. The crest gates were another of the many massive components of the dam. Each crest gate measured 22.5 feet high by 41 feet wide and weighed 42 tons. The gates were designed to maintain the elevation of the reservoir behind the dam and, by opening and closing them, to regulate water level and take full advantage of the river's power, regardless of the variation in flow. The left image shows the reinforced back of the gate; the image below shows the front of the gate that faced the reservoir.

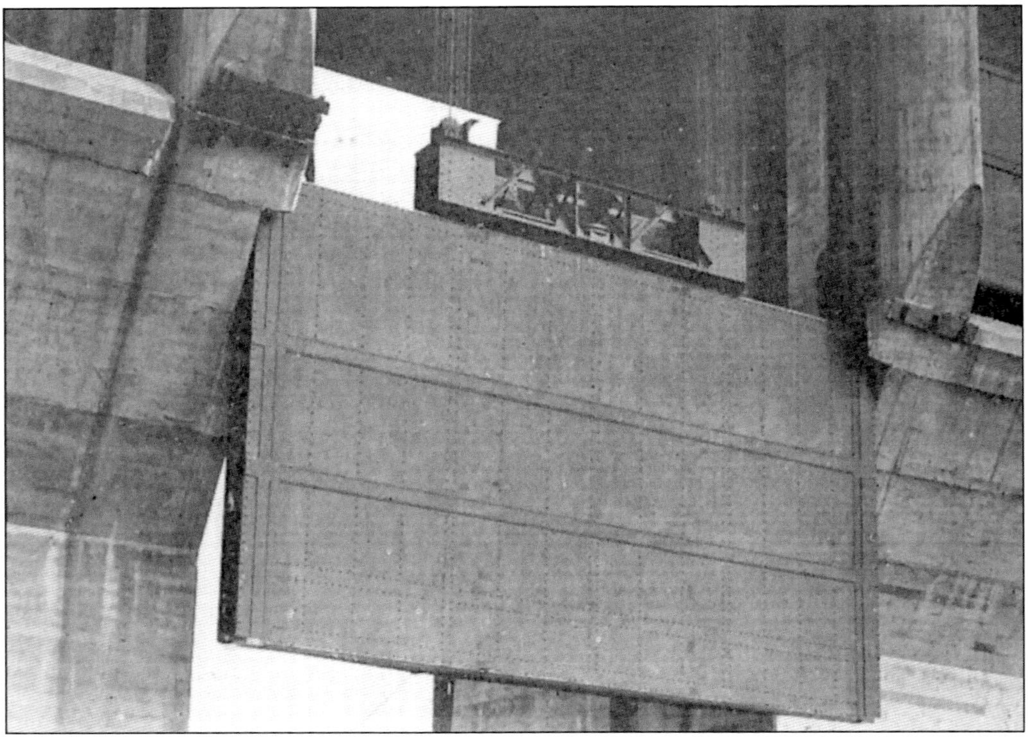

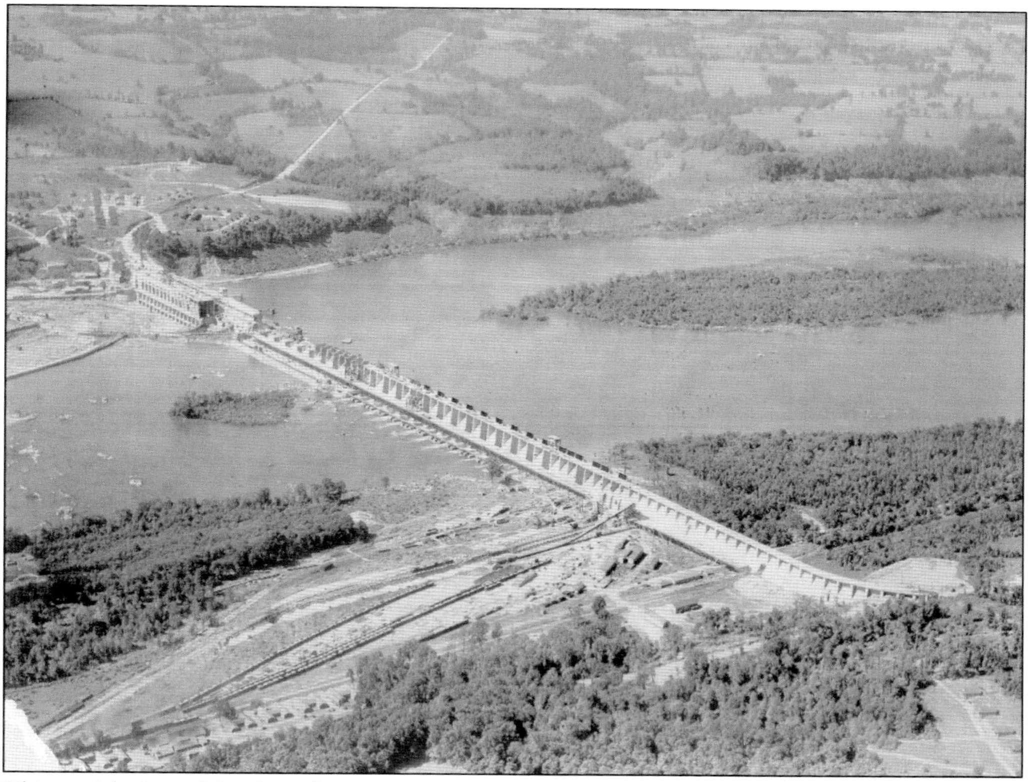

This aerial view of the dam shows the spillway nearing completion, with concrete being poured into the final middle sections of the dam to close them off. Many newly installed crest gates are also visible in their raised position along the top of the spillway. The rail lines carrying supplies upriver from Port Deposit are also prominent in the foreground. (Courtesy of the Hagley Museum.)

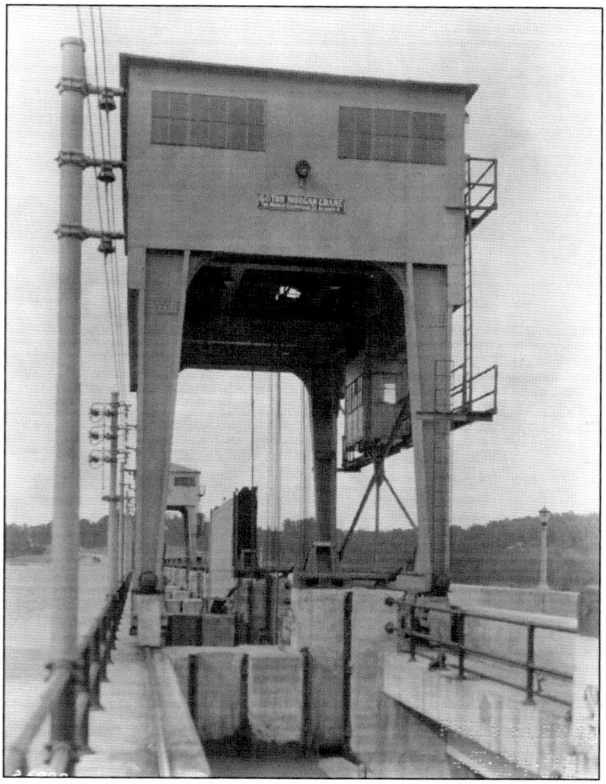

Operators raised and lowered crest gates by means of three 60-ton gantry cranes (shown here), which traveled back and forth along a track on the top of the dam at a speed of 300 feet per minute. The cranes lifted the gates one at a time with a hoisting speed of 10 feet per minute.

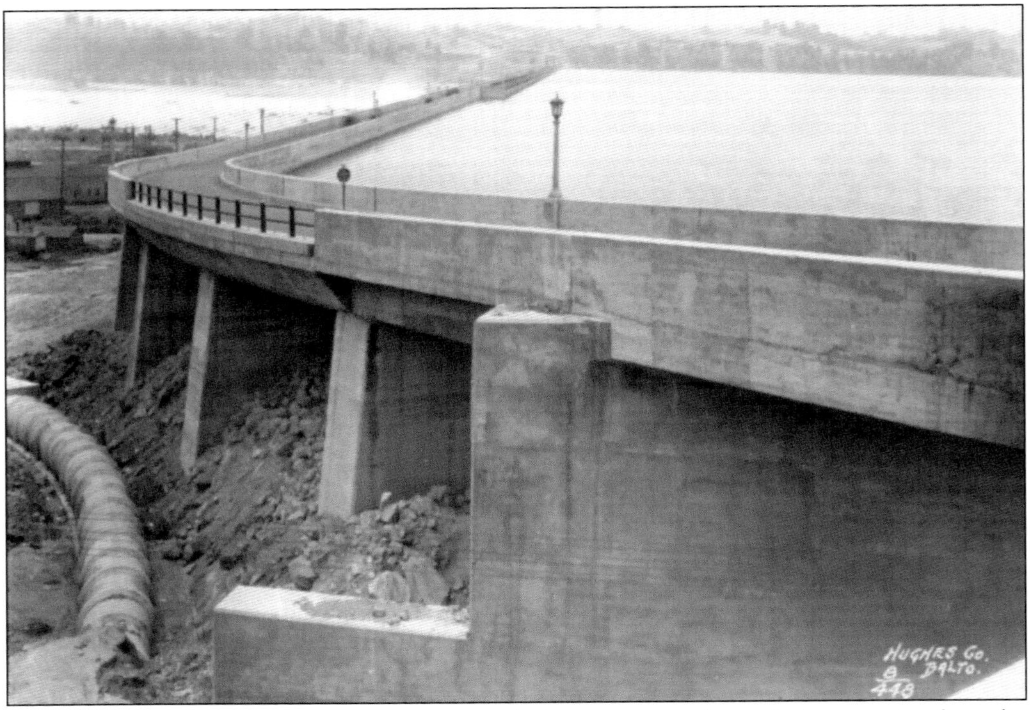

An essential aspect of the project was rerouting the important north-south river crossing from the town of Conowingo, Maryland, to the top of the dam. The roadway was designed and owned by the Maryland State Highway Administration, which also maintained it. The photograph above is of the graceful curve that marked the highway's approach from the Cecil County shore onto the bridge. The view below is from the center of the dam looking east toward Cecil County, Maryland. Visible on the left of the roadway is a single crest gate in the raised, or open, position. When the dam was in spill conditions with crest gates open, water ran under the roadway and through the opening provided by the raised gates.

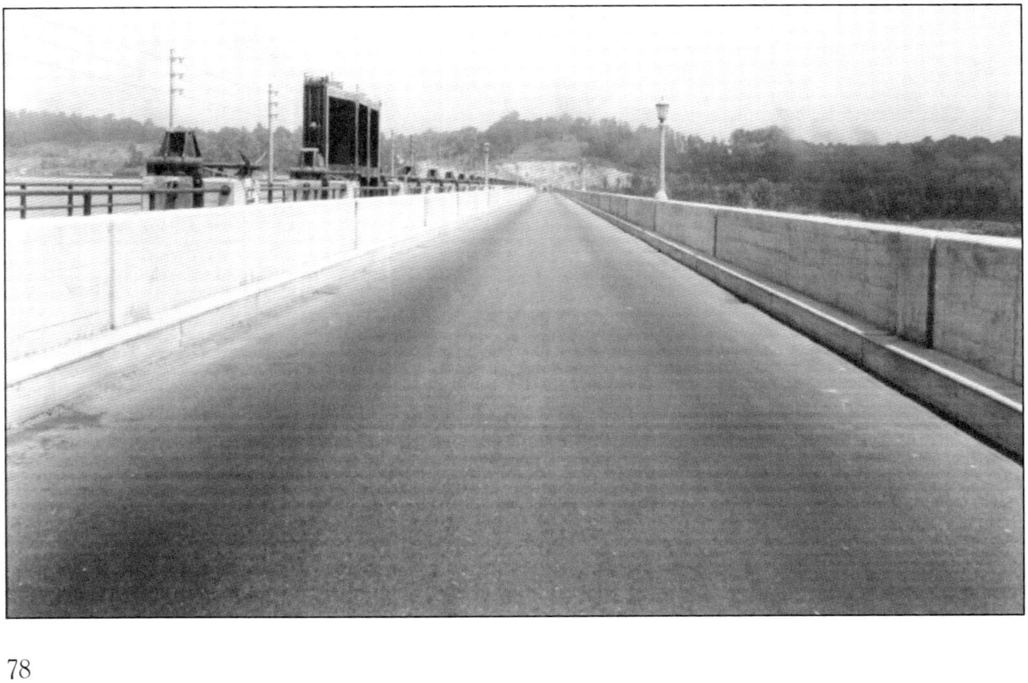

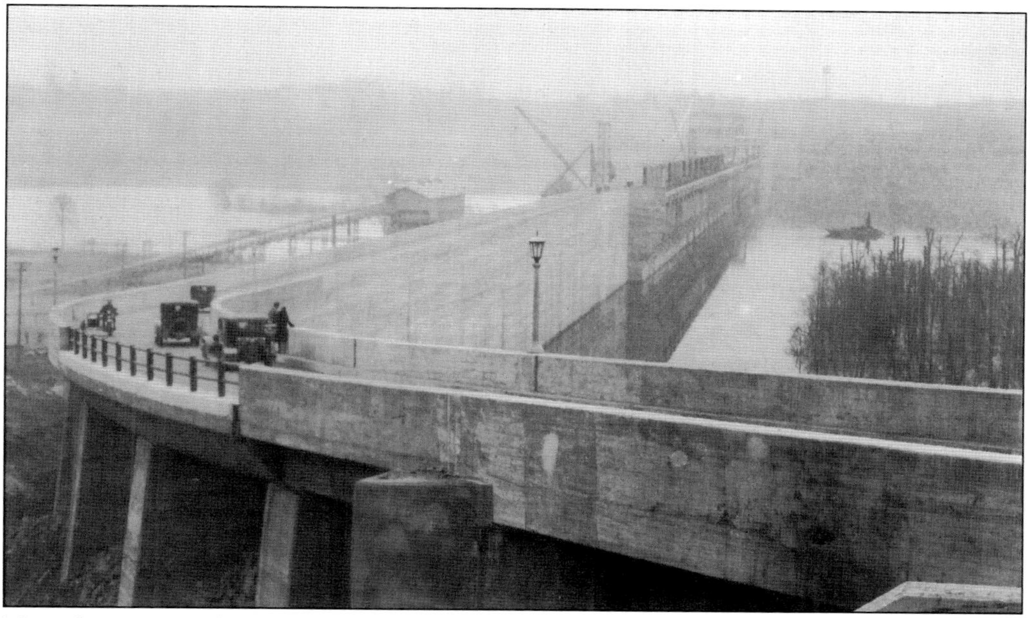

More than 18 months into the two-year construction project, the Conowingo Dam was taking its final shape. Work was moving along so well that the new roadway over its crest opened to the public nine months ahead of schedule. At one o'clock in the afternoon on Tuesday, November 15, 1927, a procession of automobiles carrying dignitaries from the Arundel Corporation and the Philadelphia Electric Company made the inaugural half-mile crossing. By then, the dam had gained some fame and the public's attention was piqued. On the first Sunday of operation, the new crossing attracted nearly 5,000 automobiles filled with curious sightseers anxious to motor over the Susquehanna at the majestic height of 120 feet. While they marveled and gawked, work to finish the dam continued apace. The formal opening of the dam was less than four months away. (Above, courtesy of the Cecil County Historical Society; below, courtesy of the *Baltimore Sun*.)

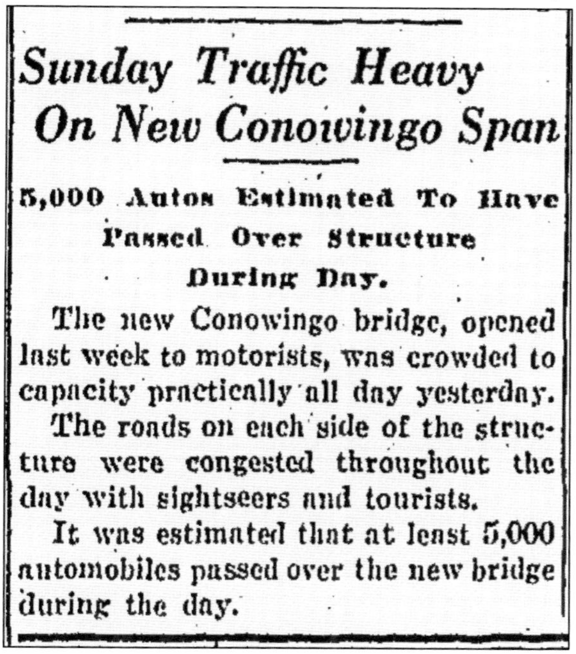

Sunday Traffic Heavy On New Conowingo Span

5,000 Autos Estimated To Have Passed Over Structure During Day.

The new Conowingo bridge, opened last week to motorists, was crowded to capacity practically all day yesterday.

The roads on each side of the structure were congested throughout the day with sightseers and tourists.

It was estimated that at least 5,000 automobiles passed over the new bridge during the day.

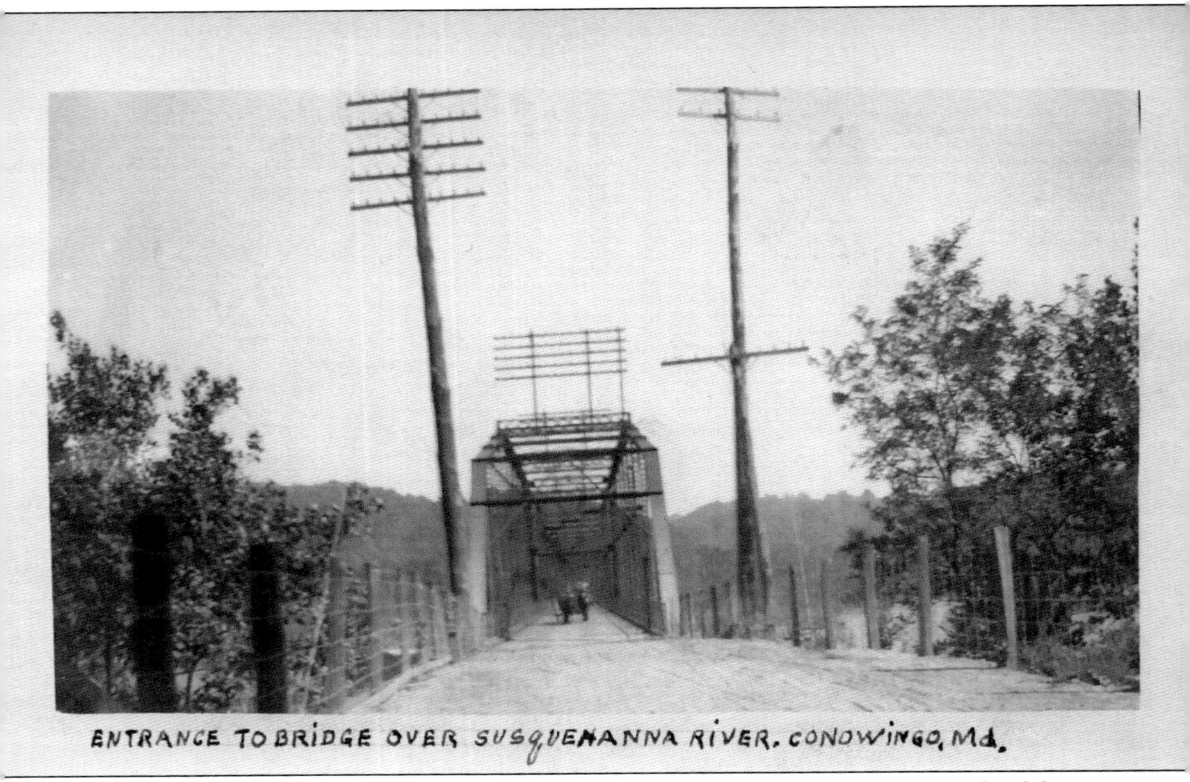

When the new bridge opened over the dam, the old crossing at Conowingo, Maryland, became obsolete. A bridge was first erected there by the Rock Run Bridge and Banking Company in 1818. It was taken out by a stampeding herd of cattle in 1854, and then a few years later, the repaired bridge was swept away by an unprecedented freshet, the sudden surge of water caused by the annual spring thaws that rush down the Susquehanna River watershed. A new enterprise, the Conowingo Bridge Company, erected a replacement in 1859, but the eastern span of that bridge was swept away by another freshet in 1904, and then the western half was destroyed by fire in 1907. The whole structure was finally remade with steel, as seen in this photograph. (Courtesy of the Cecil County Historical Society.)

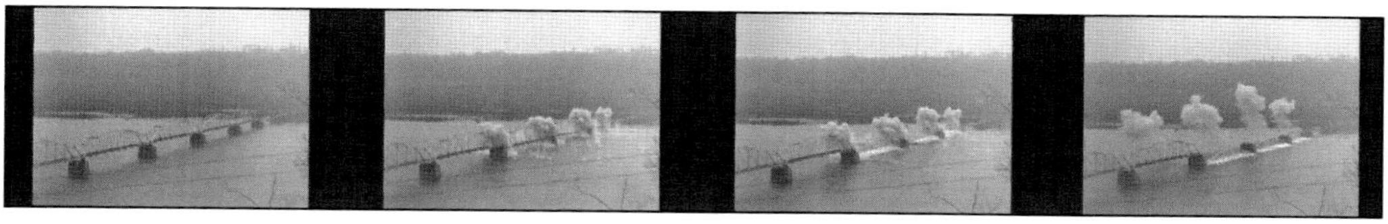

On a chilly November 28, 1927, a cameraman of the US Army Signal Corps turned his lens on the old Conowingo Bridge and captured its demise. The sequential frames from that newsreel show the thunderous moment when the bridge was blown from its stone foundations and the steel span slumped into the river. The blast was front-page news in Baltimore as readers of the *Sun* newspaper absorbed the spectacle: "The old Conowingo Bridge is no more. Forty-eight charges of nitro-gelatin set off in two explosions yesterday snapped its main beams like toothpicks, and last night the seven twisted spans of the old steel structure lay at the bottom of the Susquehanna." According to demolition engineers, the explosives were perfectly placed and the bridge fell just as they had planned. (Above, author's collection; below, courtesy of the *Baltimore Sun*.)

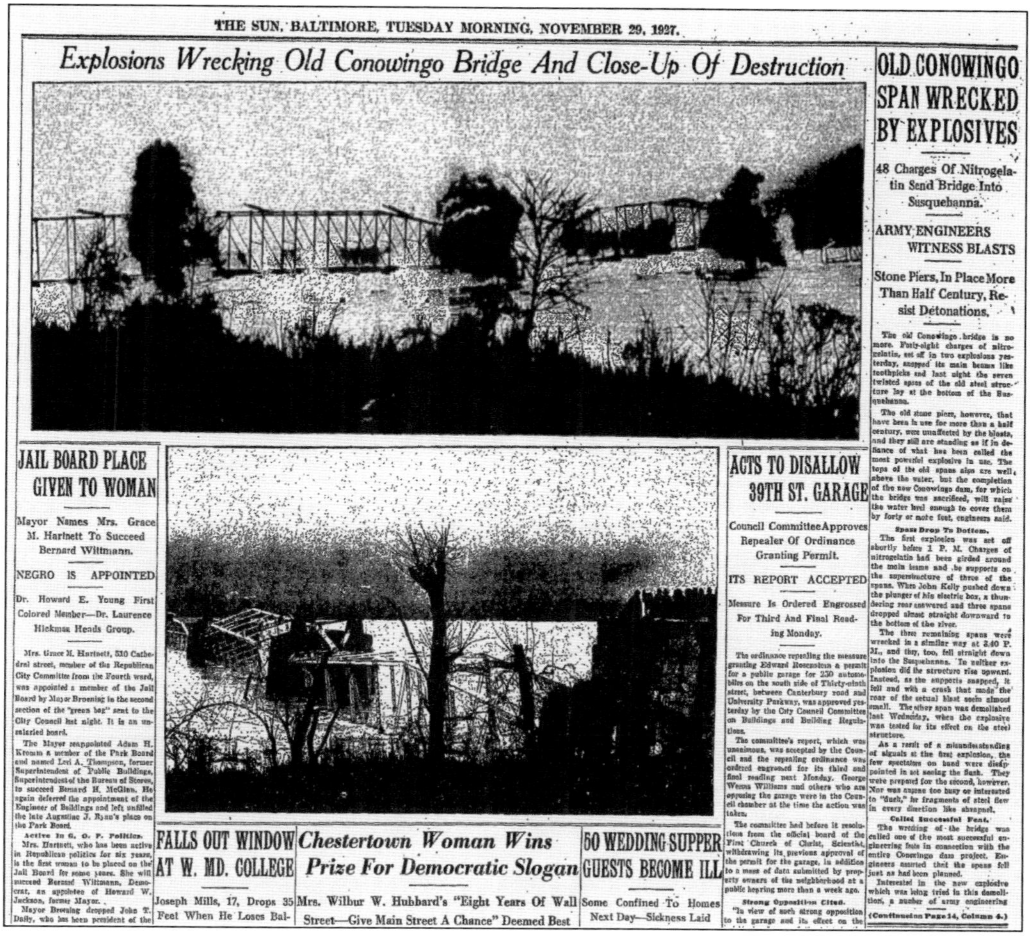

THE SUN, BALTIMORE, TUESDAY MORNING, NOVEMBER 29, 1927.

Explosions Wrecking Old Conowingo Bridge And Close-Up Of Destruction

OLD CONOWINGO SPAN WRECKED BY EXPLOSIVES

48 Charges Of Nitrogelatin Send Bridge Into Susquehanna.

ARMY ENGINEERS WITNESS BLASTS

Stone Piers, In Place More Than Half Century, Resist Detonations.

The old Conowingo bridge is no more. Forty-eight charges of nitrogelatin, set off in two explosions yesterday, snapped its main beams like toothpicks and last night the seven twisted spans of the old steel structure lay at the bottom of the Susquehanna.

The old stone piers, however, that have been in use for more than a half century, were unaffected by the blasts, and they still are standing as if in defiance of what has been called the most powerful explosive in use. The tops of the old spans also are well above the water, but the completion of the new Conowingo dam, for which the bridge was sacrificed, will raise the water level enough to cover them by forty or more feet, engineers said.

Spans Drop To Bottom.

The first explosion was set off shortly before 1 P. M. Charges of nitrogelatin had been girded around the main beams and the supports on the superstructure of three of the spans. When John Kelly pushed down the plunger of his electric box, a thundering roar answered and three spans dropped almost straight downward to the bottom of the river.

The three remaining spans were wrecked in a similar way at 3.40 P. M., and they, too, fell straight down into the Susquehanna. In neither explosion did the structure rise upward. Instead, as the supports snapped, it fell and with a crash that made the roar of the actual blast seem almost small. The other span was demolished last Wednesday, when the explosive was tested for its effect on the steel structure.

As a result of a misunderstanding of signals at the first explosion, the few spectators on hand were disappointed in not seeing the flash. They were prepared for the second, however. Nor was anyone too busy or interested to "duck," for fragments of steel flew in every direction like shrapnel.

Called Successful Feat.

The wrecking of the bridge was called one of the most successful engineering feats in connection with the entire Conowingo dam project. Engineers asserted that the spans fell just as had been planned.

Interested in the new explosive which was being tried in this demolition, a number of army engineering

(Continued on Page 14, Column 4.)

JAIL BOARD PLACE GIVEN TO WOMAN

Mayor Names Mrs. Grace M. Hartnett To Succeed Bernard Wittmann.

NEGRO IS APPOINTED

Dr. Howard E. Young First Colored Member—Dr. Laurence Hickman Heads Group.

Mrs. Grace M. Hartnett, 510 Cathedral street, member of the Republican City Committee from the Fourth ward, was appointed a member of the Jail Board by Mayor Broening in the second section of the "green bag" sent to the City Council last night. It is an unsalaried board.

The Mayor reappointed Adam H. Kremm a member of the Park Board and named Levi A. Thompson, former Superintendent of Public Buildings, Superintendent of the Bureau of Stores, to succeed Bernard H. McGinn. He again deferred the appointment of the Engineer of Buildings and left unfilled the late Augustine J. Ryan's place on the Park Board.

Active In G. O. P. Politics.

Mrs. Hartnett, who has been active in Republican politics for six years, is the first woman to be placed on the Jail Board for some years. She will succeed Bernard Wittmann, Democrat, an appointee of Howard W. Jackson, former Mayor.

Mayor Broening dropped John T. Daily, who has been president of the

ACTS TO DISALLOW 39TH ST. GARAGE

Council Committee Approves Repealer Of Ordinance Granting Permit.

ITS REPORT ACCEPTED

Measure Is Ordered Engrossed For Third And Final Reading Monday.

The ordinance repealing the measure granting Edward Reseastein a permit for a public garage for 230 automobiles on the south side of Thirty-ninth street, between Canterbury road and University Parkway, was approved yesterday by the City Council Committee on Buildings and Building Regulations.

The committee's report, which was unanimous, was accepted by the Council and the repealing ordinance was ordered engrossed for its third and final reading next Monday. George Weems Williams and others who are opposing the garage were in the Council chamber at the time the action was taken.

The committee had before it resolutions from the official board of the First Church of Christ, Scientist, withdrawing its previous approval of the permit for the garage, in addition to a mass of data submitted by property owners of the neighborhood at a public hearing more than a week ago.

Strong Opposition Cited.

"In view of such strong opposition to the garage and its effect on the

FALLS OUT WINDOW AT W. MD. COLLEGE

Joseph Mills, 17, Drops 35 Feet When He Loses Bal-

Chestertown Woman Wins Prize For Democratic Slogan

Mrs. Wilbur W. Hubbard's "Eight Years Of Wall Street—Give Main Street A Chance" Deemed Best

50 WEDDING SUPPER GUESTS BECOME ILL

Some Confined To Homes Next Day—Sickness Laid

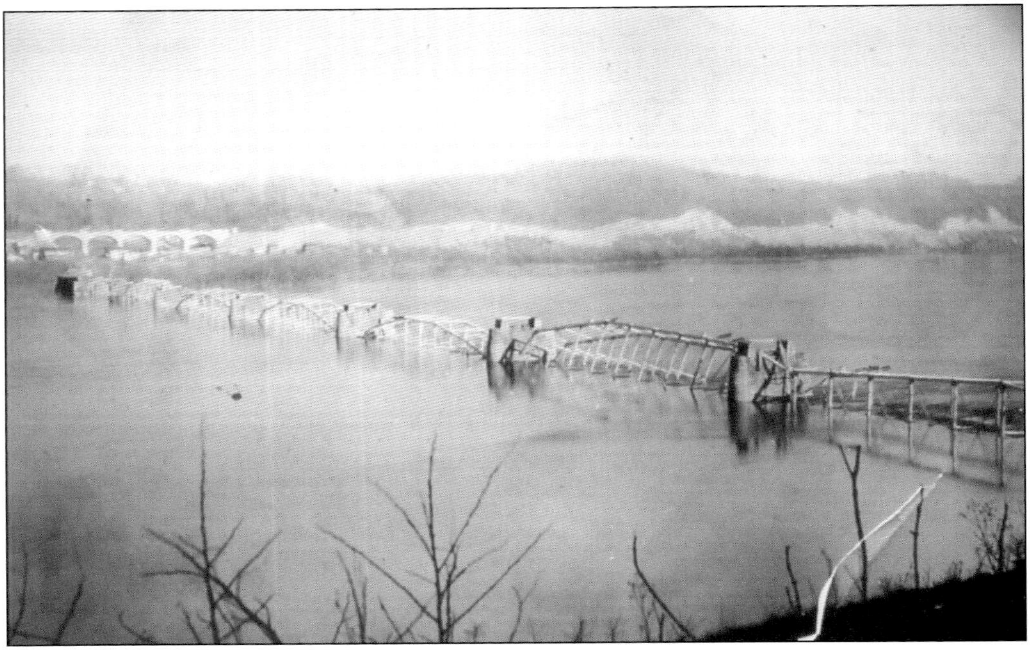

After the explosions, the remnants of the old Conowingo Bridge rested in the Susquehanna shallows. Presumably, salvagers arrived, hauled away the twisted wreckage, and sold the valuable scrap steel, but they left behind the stone foundations. "The old stone piers," a *Baltimore Sun* reporter embellished, "were unaffected by the blasts, and they are standing as if in defiance of what has been called the most powerful explosive in use . . . but the completion of the new Conowingo Dam, for which the bridge was sacrificed, will raise the water level enough to cover them by forty or more feet." When the dam opened, the piers were submerged, and they now lie at the bottom of the pond behind the dam. (Both, courtesy of the Harford County Historical Society.)

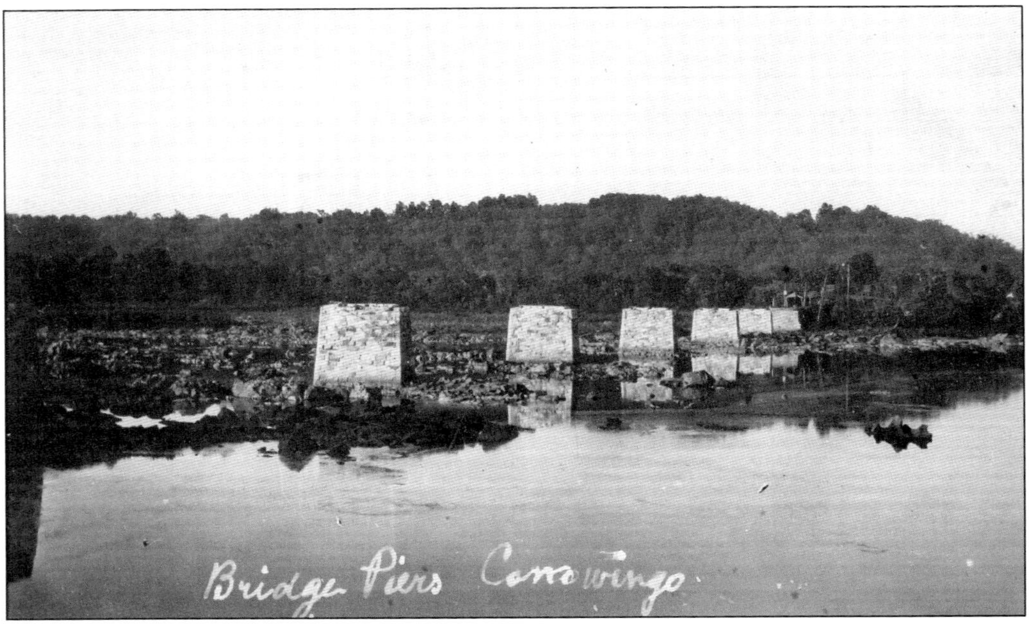

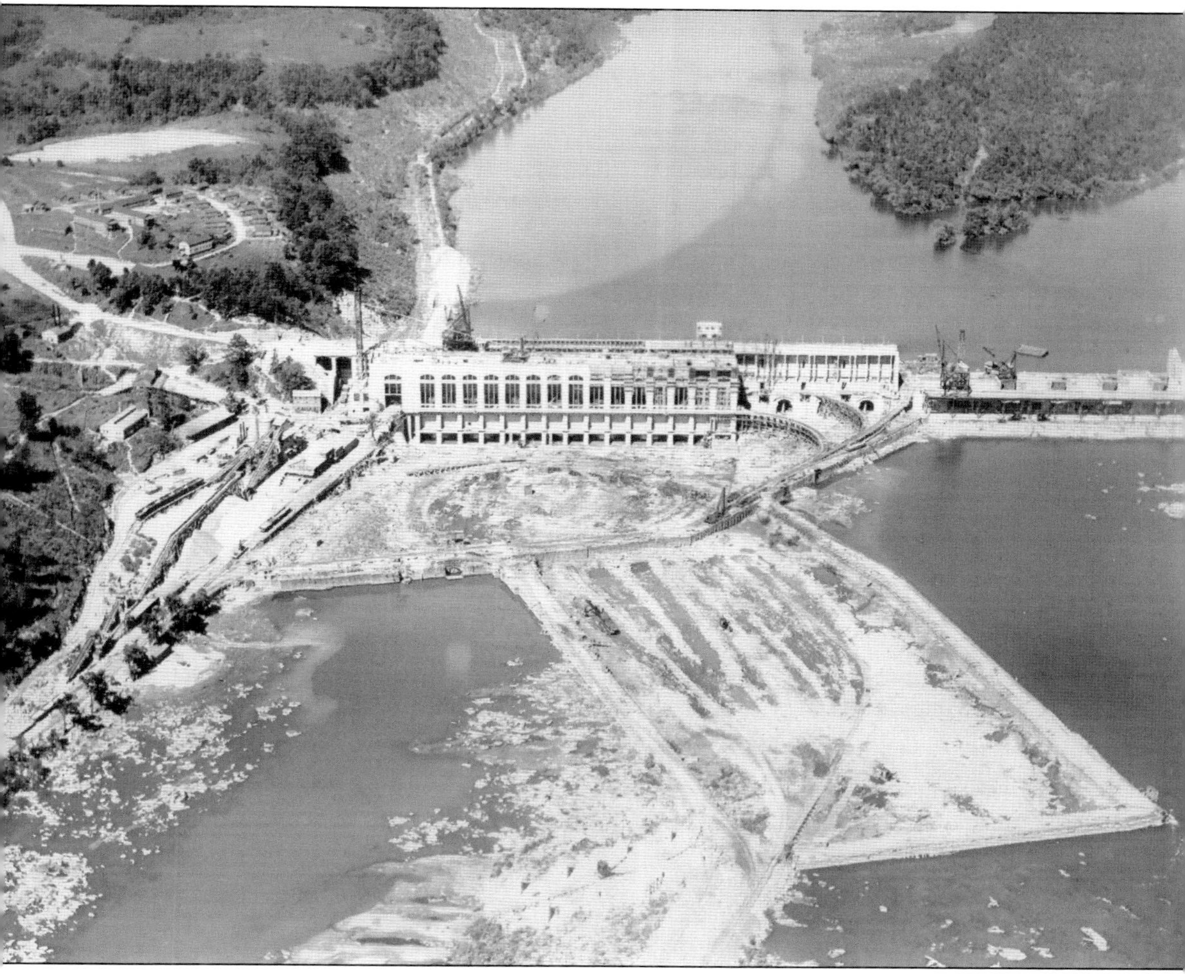

In the aerial photograph above from the fall of 1927, the finishing touches are being put on Conowingo Dam. Concrete work on the powerhouse, all 235,917 cubic yards of it, was completed on December 31, 1927. A few weeks later, in mid-January, the final loads of concrete were chuted into the last openings of the spillway, completing the pouring of a whopping 428,000 cubic yards of concrete. In the photograph below, all concrete towers and chutes have been removed, save one, and the temporary rail lines that once ran supplies over the cofferdams have been torn up. Work to erect the transformer and transmission station on the roof of the powerhouse is about to begin. (Above, courtesy of the Hagley Museum.)

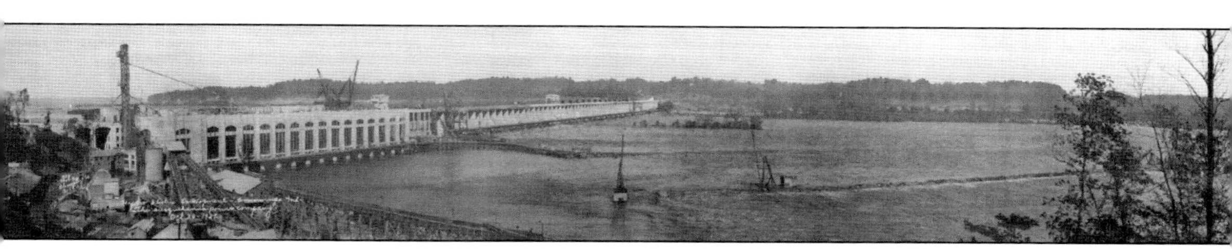

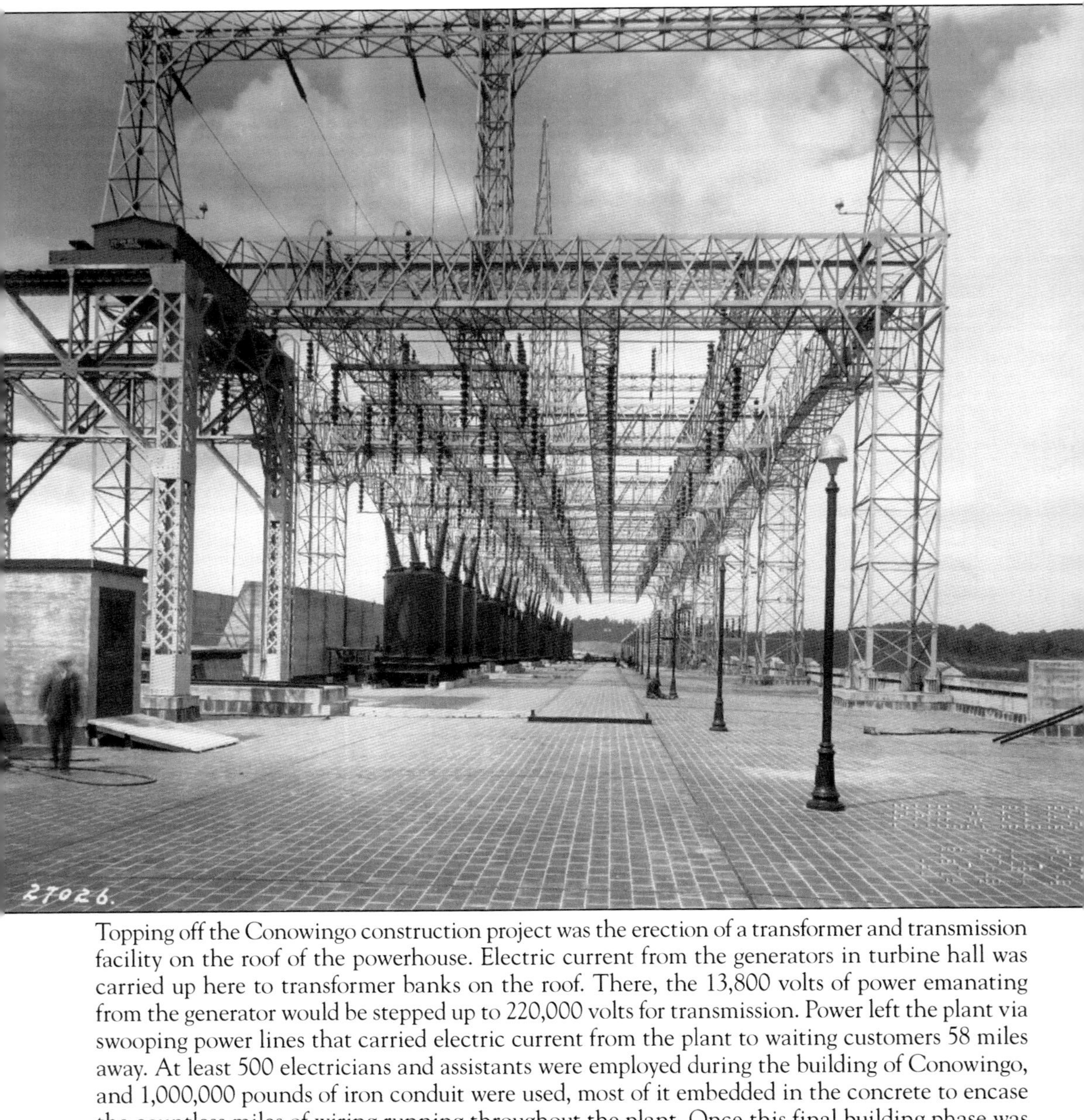

Topping off the Conowingo construction project was the erection of a transformer and transmission facility on the roof of the powerhouse. Electric current from the generators in turbine hall was carried up here to transformer banks on the roof. There, the 13,800 volts of power emanating from the generator would be stepped up to 220,000 volts for transmission. Power left the plant via swooping power lines that carried electric current from the plant to waiting customers 58 miles away. At least 500 electricians and assistants were employed during the building of Conowingo, and 1,000,000 pounds of iron conduit were used, most of it embedded in the concrete to encase the countless miles of wiring running throughout the plant. Once this final building phase was completed, the Conowingo Dam would be ready for the generators to be turned over and for the plant to come on line.

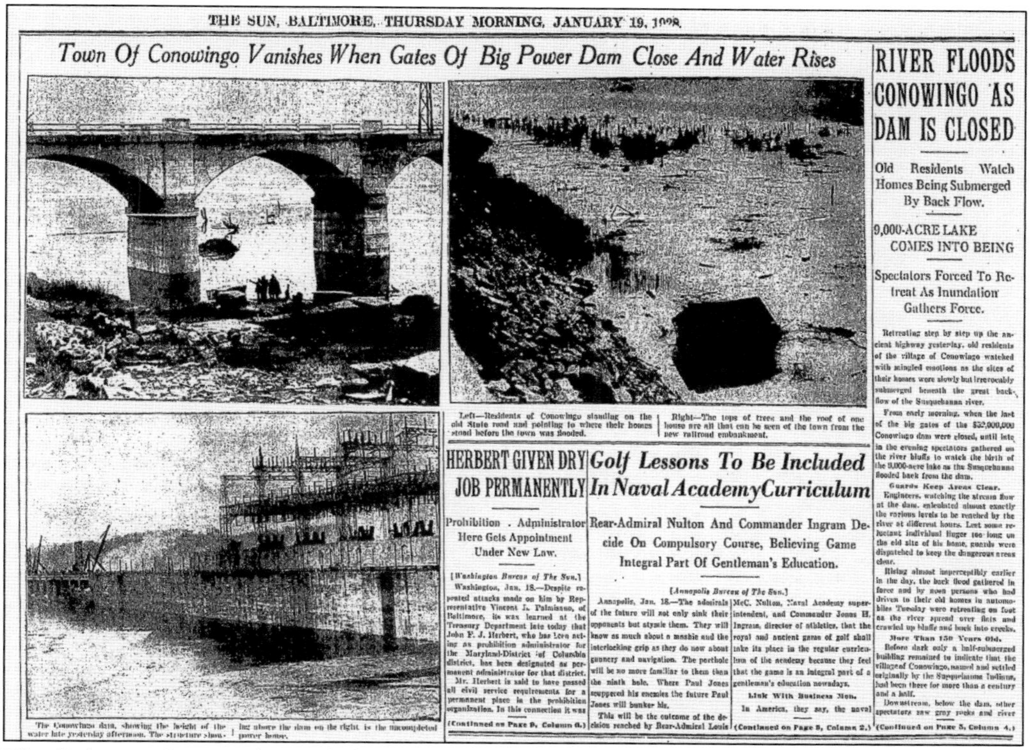

THE SUN, BALTIMORE, THURSDAY MORNING, JANUARY 19, 1928.

Town Of Conowingo Vanishes When Gates Of Big Power Dam Close And Water Rises

Left—Residents of Conowingo standing on the old State road and pointing to where their homes stood before the town was flooded.

Right—The tops of trees and the roof of one house are all that can be seen of the town from the new railroad embankment.

The Conowingo dam, showing the height of the water late yesterday afternoon. The structure showing above the dam on the right is the uncompleted power house.

RIVER FLOODS CONOWINGO AS DAM IS CLOSED

Old Residents Watch Homes Being Submerged By Back Flow.

9,000-ACRE LAKE COMES INTO BEING

Spectators Forced To Retreat As Inundation Gathers Force.

Retreating step by step up the ancient highway yesterday, old residents of the village of Conowingo watched with mingled emotions as the sites of their homes were slowly but irrevocably submerged beneath the great back-flow of the Susquehanna river.

From early morning, when the last of the big gates of the $52,000,000 Conowingo dam were closed, until late in the evening spectators gathered on the river bluffs to watch the birth of the 9,000-acre lake as the Susquehanna flooded back from the dam.

Guards Keep Areas Clear.

Engineers, watching the stream flow at the dam, calculated almost exactly the various levels to be reached by the river at different hours. Lest some reluctant individual linger too long on the old site of his home, guards were dispatched to keep the dangerous areas clear.

Rising almost imperceptibly earlier in the day, the back flood gathered in force and by noon persons who had driven to their old homes in automobiles Tuesday were retreating on foot as the river spread over flats and crawled up bluffs and back into creeks.

More Than 150 Years Old.

Before dark only a half-submerged building remained to indicate that the village of Conowingo, named and settled originally by the Supquehanna Indians, had been there for more than a century and a half.

Downstream, below the dam, other spectators saw gray rocks and river

(Continued on Page 5, Column 4.)

HERBERT GIVEN DRY JOB PERMANENTLY

Prohibition Administrator Here Gets Appointment Under New Law.

[*Washington Bureau of The Sun.*]

Washington, Jan. 18.—Despite repeated attacks made on him by Representative Vincent L. Palmisano, of Baltimore, it was learned at the Treasury Department late today that John F. J. Herbert, who has been acting as prohibition administrator for the Maryland-District of Columbia district, has been designated as permanent administrator for that district.

Mr. Herbert is said to have passed all civil service requirements for a permanent place in the prohibition organization. In this connection it was

(Continued on Page 9, Column 6.)

Golf Lessons To Be Included In Naval Academy Curriculum

Rear-Admiral Nulton And Commander Ingram Decide On Compulsory Course, Believing Game Integral Part Of Gentleman's Education.

[*Annapolis Bureau of The Sun.*]

Annapolis, Jan. 18.—The admirals of the future will not only sink their opponents but stymie them. They will know as much about a mashie and the interlocking grip as they do now about gunnery and navigation. The porthole will be no more familiar to them than the ninth hole. Where Paul Jones scuppered his enemies the future Paul Jones will bunker his.

This will be the outcome of the decision reached by Rear-Admiral Louis McC. Nulton, Naval Academy superintendent, and Commander Jonas H. Ingram, director of athletics, that the royal and ancient game of golf shall take its place in the regular curriculum of the academy because they feel that the game is an integral part of a gentleman's education nowadays.

Link With Business Men.

In America, they say, the naval

(Continued on Page 5, Column 2.)

This *Baltimore Sun* article is titled "Town of Conowingo Vanishes When Gates of Big Power Dam Close and Water Rises." On the cold morning of January 18, 1928, the big gates at Conowingo Dam closed. Slowly, almost imperceptibly at first, river water, which had once flowed so freely, now backed up and pooled behind the dam. By the end of the day, not more than a mile upstream, the town of Conowingo was no more. The poignancy of the occasion—water inexorably inching up the hillsides, people watching their village vanish—was not lost on the reporter from *Baltimore Sun*: "Retreating step-by-step up the ancient highway yesterday, old residents of the village of Conowingo watched with mingled emotions as the sites of their homes were slowly but irrevocably submerged beneath the great back-flow of the Susquehanna River. From early morning, when the last of the big gates of the $52,000,000 Conowingo Dam were closed, until late in the evening spectators gathered on the river bluffs to watch the birth of the 9,000 acre lake as the Susquehanna flooded back from the dam." (Courtesy of the *Baltimore Sun*.)

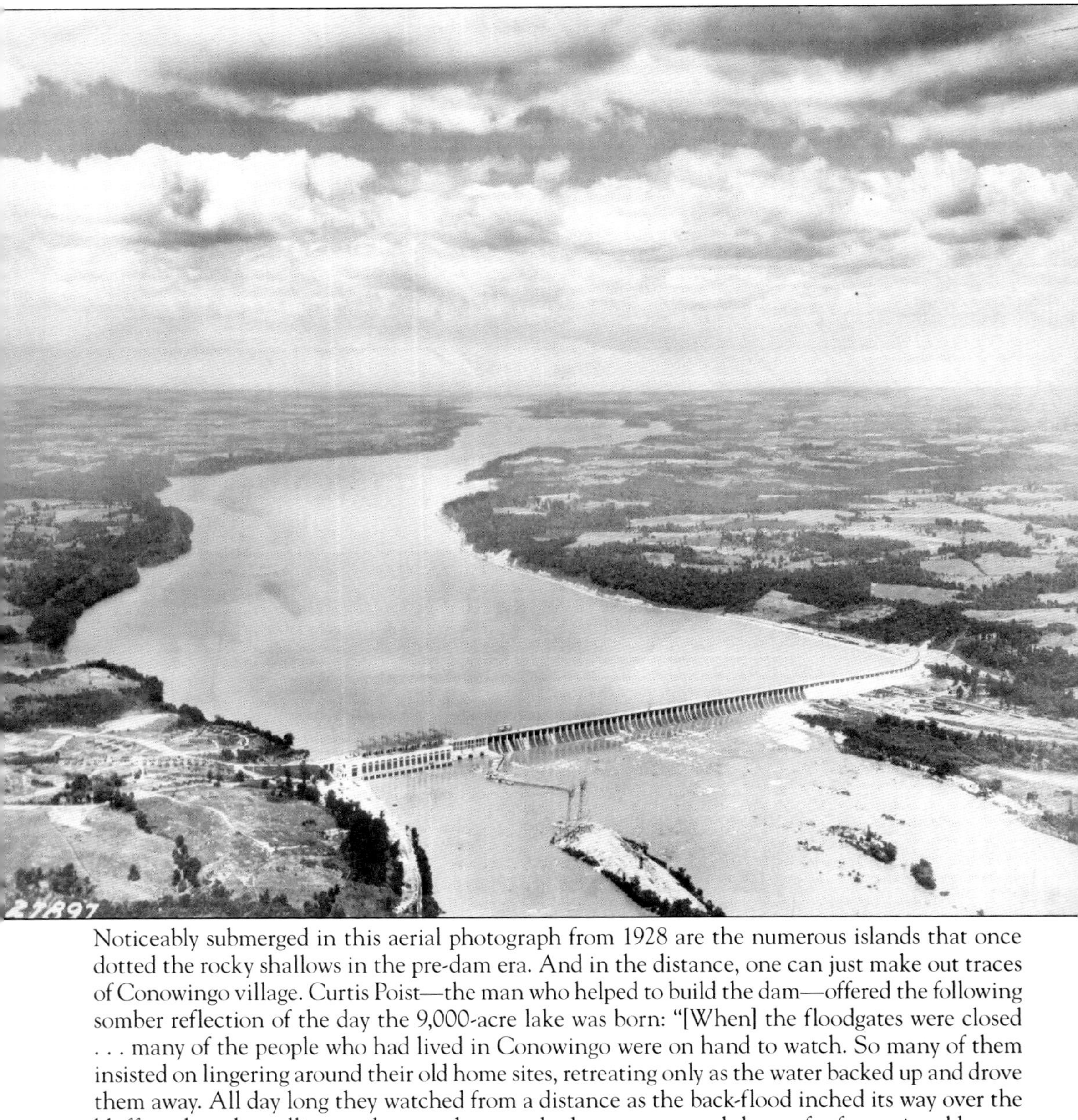

Noticeably submerged in this aerial photograph from 1928 are the numerous islands that once dotted the rocky shallows in the pre-dam era. And in the distance, one can just make out traces of Conowingo village. Curtis Poist—the man who helped to build the dam—offered the following somber reflection of the day the 9,000-acre lake was born: "[When] the floodgates were closed . . . many of the people who had lived in Conowingo were on hand to watch. So many of them insisted on lingering around their old home sites, retreating only as the water backed up and drove them away. All day long they watched from a distance as the back-flood inched its way over the bluffs and up the gullies until, at sundown, only the tree tops and the roofs of occasional houses and barns remained above water to identify the place which had once been home."

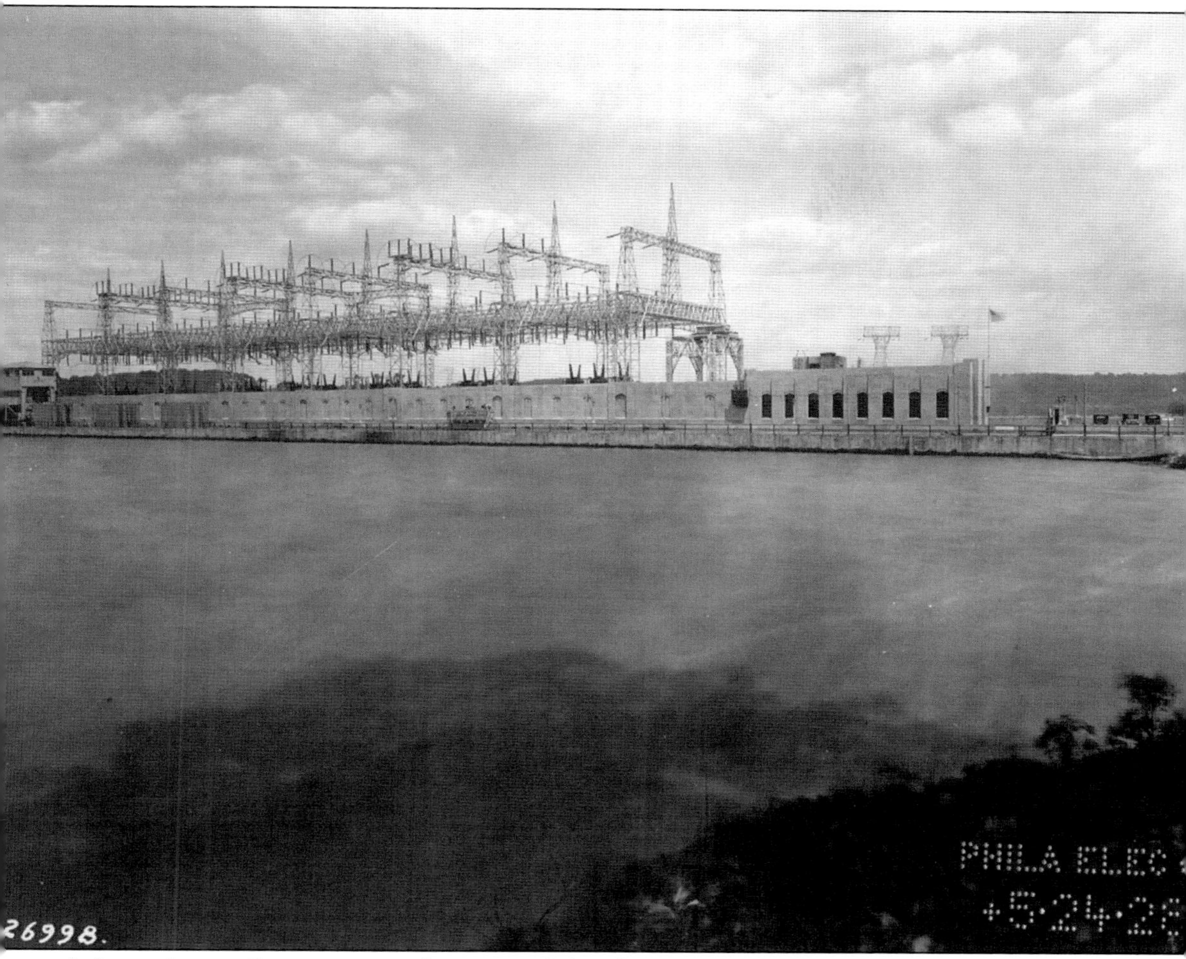

A few ambitious Conowingo residents moved their homes to a hill above the dam. One of the homes, the 2½-story Ragan House, owned by the town doctor, David Marshal "Doc" Reagan, still stands on the Cecil County side of the river. The rest of Conowingo's buildings, as well as many farmsteads and country homes in the vicinity, were abandoned and flooded. Engineers working that day said they could not determine accurately just when the water would reach the top of the dam, but it did not take long. By sundown, the river was already passing over the spillway, and the newly formed Conowingo Pond, which ranged in depth from 40 to 60 feet, stood ready to send its water through the turbines of the new dam.

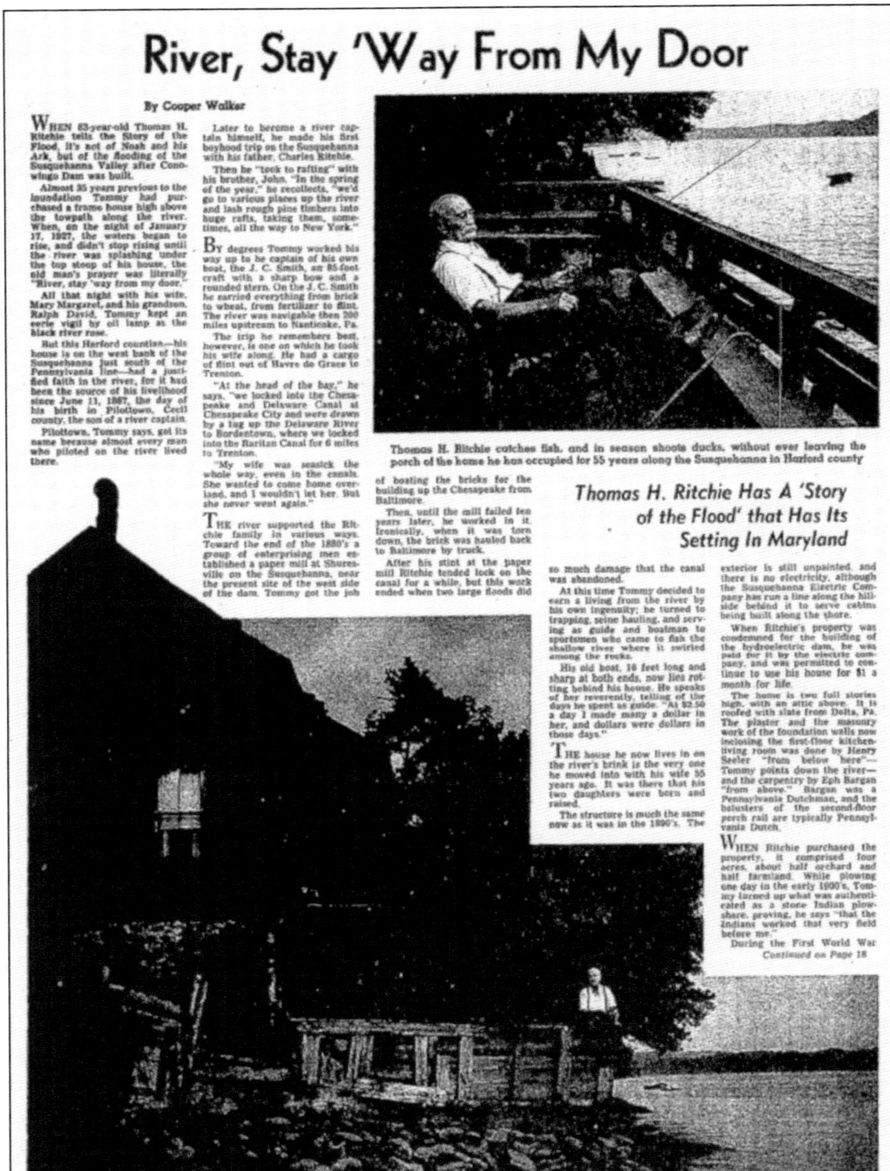

River, Stay 'Way From My Door

By Cooper Walker

WHEN 83-year-old Thomas H. Ritchie tells the Story of the Flood, it's not of Noah and his Ark, but of the flooding of the Susquehanna Valley after Conowingo Dam was built.

Almost 35 years previous to the inundation Tommy had purchased a frame house high above the towpath along the river. When, on the night of January 17, 1927, the waters began to rise, and didn't stop rising until the river was splashing under the top stoop of his house, the old man's prayer was literally "River, stay 'way from my door."

All that night with his wife, Mary Margaret, and his grandson, Ralph David, Tommy kept an eerie vigil by oil lamp as the black river rose.

But this Harford countian—his house is on the west bank of the Susquehanna just south of the Pennsylvania line—had a justified faith in the river, for it had been the source of his livelihood since June 11, 1867, the day of his birth in Pilottown, Cecil county, the son of a river captain.

Pilottown, Tommy says, got its name because almost every man who piloted on the river lived there.

Later to become a river captain himself, he made his first boyhood trip on the Susquehanna with his father, Charles Ritchie.

Then he "took to rafting" with his brother, John. "In the spring of the year," he recollects, "we'd go to various places up the river and lash rough pine timbers into huge rafts, taking them, sometimes, all the way to New York."

BY degrees Tommy worked his way up to be captain of his own boat, the J. C. Smith, an 85-foot craft with a sharp bow and a rounded stern. On the J. C. Smith he carried everything from brick to wheat, from fertilizer to flint. The river was navigable then 200 miles upstream to Nanticoke, Pa.

The trip he remembers best, however, is one on which he took his wife along. He had a cargo of flint out of Havre de Grace to Trenton.

"At the head of the bay," he says, "we locked into the Chesapeake and Delaware Canal at Chesapeake City and were drawn by a tug up the Delaware River to Bordentown, where we locked into the Raritan Canal for 6 miles to Trenton.

"My wife was seasick the whole way, even in the canals. She wanted to come home overland, and I wouldn't let her. But she never went again."

THE river supported the Ritchie family in various ways. Toward the end of the 1880's a group of enterprising men established a paper mill at Shuresville on the Susquehanna, near the present site of the west side of the dam. Tommy got the job of boating the bricks for the building up the Chesapeake from Baltimore.

Then, until the mill failed ten years later, he worked in it. Ironically, when it was torn down, the brick was hauled back to Baltimore by truck.

After his stint at the paper mill Ritchie tended lock on the canal for a while, but this work ended when two large floods did so much damage that the canal was abandoned.

At this time Tommy decided to earn a living from the river by his own ingenuity; he turned to trapping, seine hauling, and serving as guide and boatman to sportsmen who came to fish the shallow river where it swirled among the rocks.

His old boat, 16 feet long and sharp at both ends, now lies rotting behind his house. He speaks of her reverently, telling of the days he spent as guide. "At $2.50 a day I made many a dollar in her, and dollars were dollars in those days."

THE house he now lives in on the river's brink is the very one he moved into with his wife 55 years ago. It was there that his two daughters were born and raised.

The structure is much the same now as it was in the 1890's. The exterior is still unpainted, and there is no electricity, although the Susquehanna Electric Company has run a line along the hillside behind it to serve cabins being built along the shore.

When Ritchie's property was condemned for the building of the hydroelectric dam, he was paid for it by the electric company, and was permitted to continue to use his house for $1 a month for life.

The home is two full stories high, with an attic above. It is roofed with slate from Delta, Pa. The plaster and the masonry work of the foundation walls now inclosing the first-floor kitchen-living room was done by Henry Seeler "from below here"—Tommy points down the river—and the carpentry by Eph Bargan "from above." Bargan was a Pennsylvania Dutchman, and the balusters of the second-floor porch rail are typically Pennsylvania Dutch.

WHEN Ritchie purchased the property, it comprised four acres, about half orchard and half farmland. While plowing one day in the early 1900's, Tommy turned up what was authenticated as a stone Indian plowshare, proving, he says "that the Indians worked that very field before me."

During the First World War

Continued on Page 18

Thomas H. Ritchie catches fish, and in season shoots ducks, without ever leaving the porch of the home he has occupied for 55 years along the Susquehanna in Harford county

Thomas H. Ritchie Has A 'Story of the Flood' that Has Its Setting In Maryland

The flooding of the valley by the Conowingo Dam brought the water to Mr. Ritchie's door, but it could not make the riverman leave his home place.

7

Before water had actually backed up behind the dam, engineers surveyed the valley and painted white marks on trees and rocks to represent the future waterline. One of these marks fell right on the doorstep of Maryland resident Thomas H. Ritchie, who was warned that his rustic house on the west bank of the Susquehanna was doomed. When the fateful day arrived, Ritchie watched as the water crept higher and higher. "I went to the garage and got my automobile out," he remembered, "The water was already to the running boards . . . my wife, Mary Margaret, packed up everything and said, 'Tom let's get out of here,' but I just kept saying, 'No; first let's just see what she does.' Well, she stopped right under the top stoop, and I said to Mary Margaret, 'We've been here a long time; I guess we'll stay." From that day forth, Ritchie could go fishing without ever having to leave the porch of his now waterfront home. (Courtesy of the *Baltimore Sun*.)

Six

The Big Dam Era

Power dams like Conowingo, which generated electric current from a clean renewable resource, were shining examples of an idea that took hold of the nation with President Roosevelt's New Deal; it was a patriotic duty to harness nature for the public good. Conowingo's generators came on line at the crescendo of the Roaring Twenties, on the cusp of the stock market crash of 1929, and as a harbinger of the massive public-works dam projects that would follow. The spread of electrification was helping to modernize society, and the conquest of America's waterways to create that power was considered a fundamental right. After the stock market crash of October 1929, when the nation spiraled down into the deepest and longest-lasting economic downturn in its history, very large-scale projects like at Conowingo also came to be seen as much more than power generators. Building these big dams was a reliable way to help pull the economy out of its malaise and get people back to work. And so it was that, in the 1930s and 1940s, hydroelectric construction projects like Fort Peck on the Missouri River, the Grand Coulee and Bonneville dams in the Pacific Northwest, the Tennessee Valley Authority, and the great colossus of Hoover Dam were on the forefront of the nation's path to recovery. Preceding them all was Conowingo Dam, a project that was privately financed—the government did not invest a dime—and was completed in less than two years, ahead of schedule and on budget.

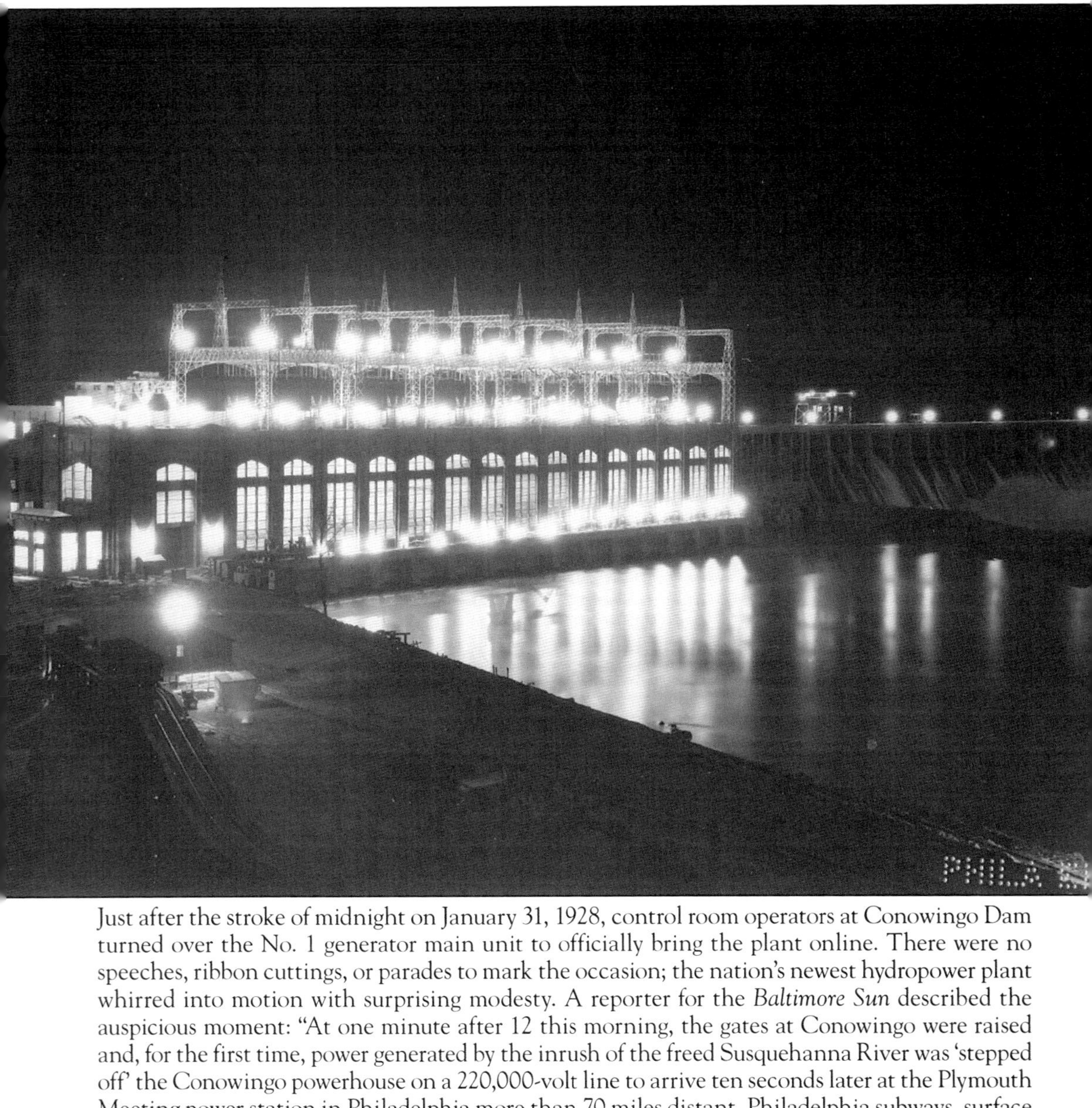

Just after the stroke of midnight on January 31, 1928, control room operators at Conowingo Dam turned over the No. 1 generator main unit to officially bring the plant online. There were no speeches, ribbon cuttings, or parades to mark the occasion; the nation's newest hydropower plant whirred into motion with surprising modesty. A reporter for the *Baltimore Sun* described the auspicious moment: "At one minute after 12 this morning, the gates at Conowingo were raised and, for the first time, power generated by the inrush of the freed Susquehanna River was 'stepped off' the Conowingo powerhouse on a 220,000-volt line to arrive ten seconds later at the Plymouth Meeting power station in Philadelphia more than 70 miles distant. Philadelphia subways, surface cars, electric light units, and all other users of electric power in that city were receiving energy from the harnessed Susquehanna River in Maryland."

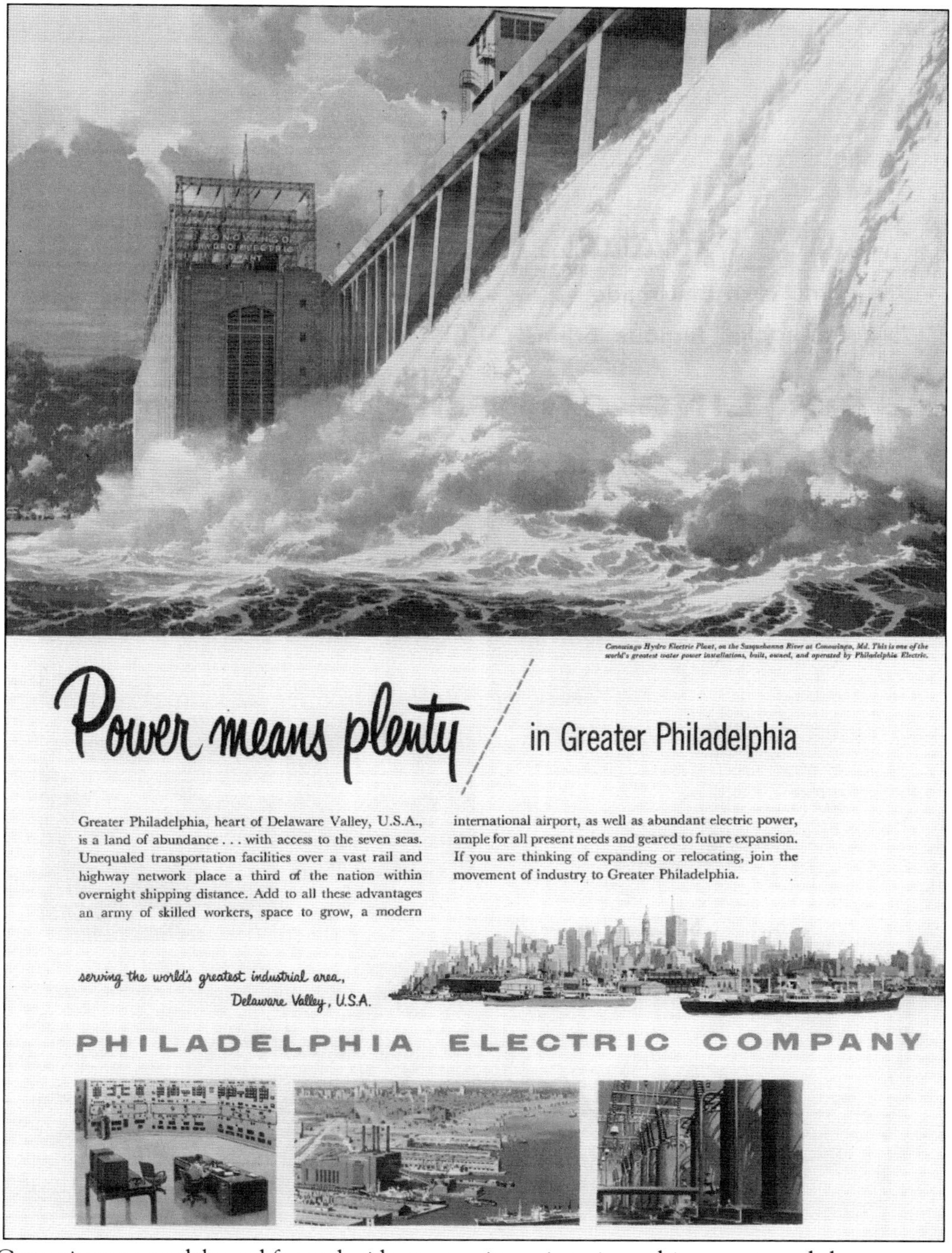

Conowingo was celebrated far and wide as an epic engineering achievement, and the company responsible for building it, the Philadelphia Electric Company, took out this advertisement, "Power Means Plenty," to promote "one of the world's greatest water power installations, built, owned, and operated by Philadelphia Electric." The advertisement presents an idealized depiction of the dam, a low-angle view of the towering structure standing strong against the river that raged through it. The message was undeniable—the Conowingo Dam had indeed harnessed the power of the Susquehanna River. The advertisement also implied that clean, renewable hydroelectric power was a guarantee of future "plenty."

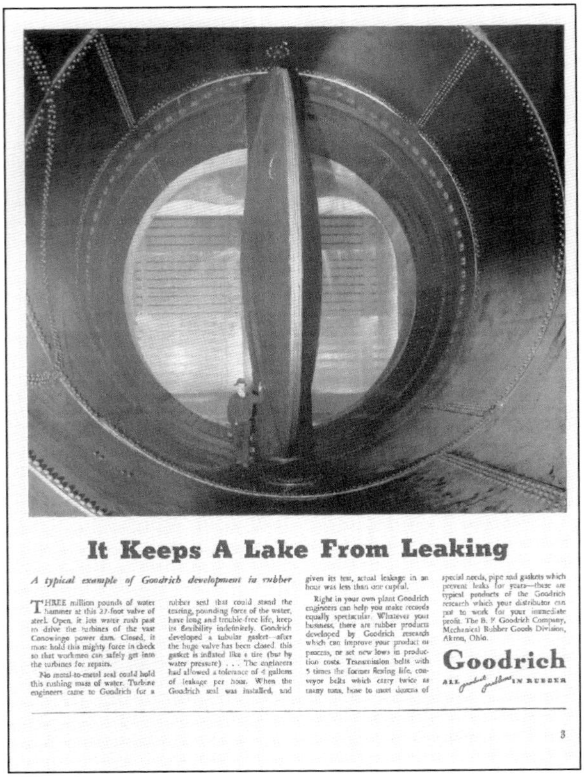

Many other companies joined in celebrating the new Conowingo Dam. In its advertisement, "It Keeps a Lake from Leaking," the B.F. Goodrich Company touts the butterfly valves it manufactured for Conowingo as follows: "Three million pounds of water hammer at this 27-foot valve of steel. Open, it lets water rush past to drive the turbines of the vast Conowingo power dam. Closed, it must hold this mighty force in check so that workmen can safely get into the turbines for repair." Likewise, the General Electric Company, which built four of the seven generating units, trumpeted its contribution in *Electrical World* magazine, noting that its generators were made with "radically new features, including welded and fabricated steel construction" and went into operation "three months ahead of schedule."

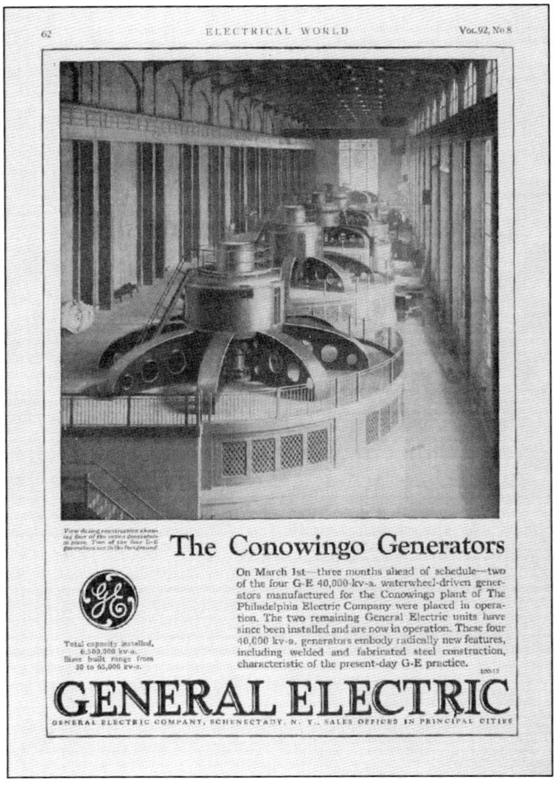

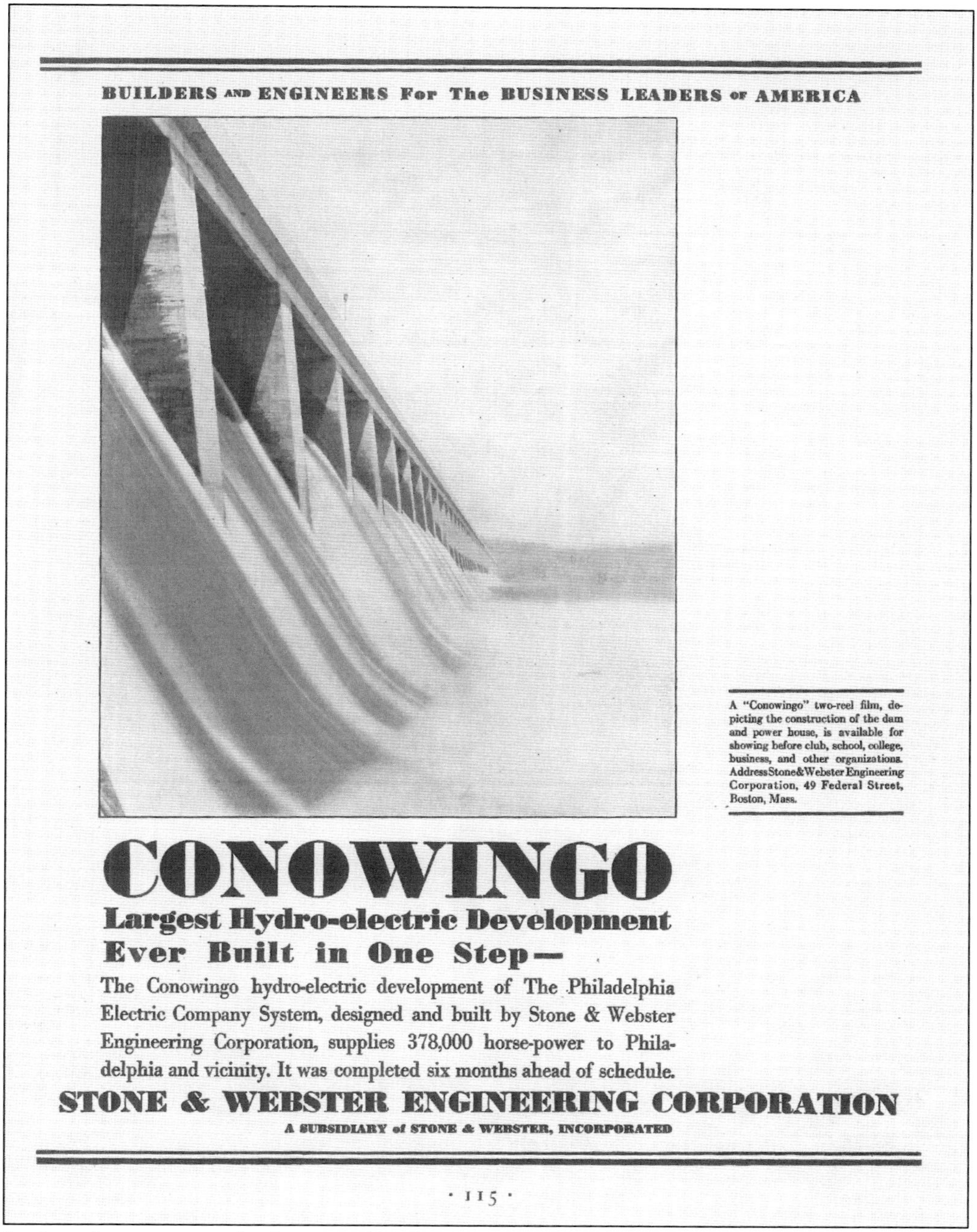

In this advertisement, the Stone and Webster Corporation, which built the dam, brands itself as "Builders and Engineers for the Business Leaders of America" and proudly notes that Conowingo Dam was the "Largest Hydro-electric Development Ever Built in One Step." The view of Conowingo, with the spillway stretching to infinity and water flowing diffusely down, looks as if rendered by an Impressionist painter. Such advertisements not only trumpet the company's accomplishments, they also romanticize the dam and convey the feeling that transforming the power of the nation's waterways into electricity is the natural course of progress. This philosophy would, in several years' time, be amplified by Pres. Franklin D. Roosevelt, who considered hydro dams instrumental for the prosperity and security of the nation.

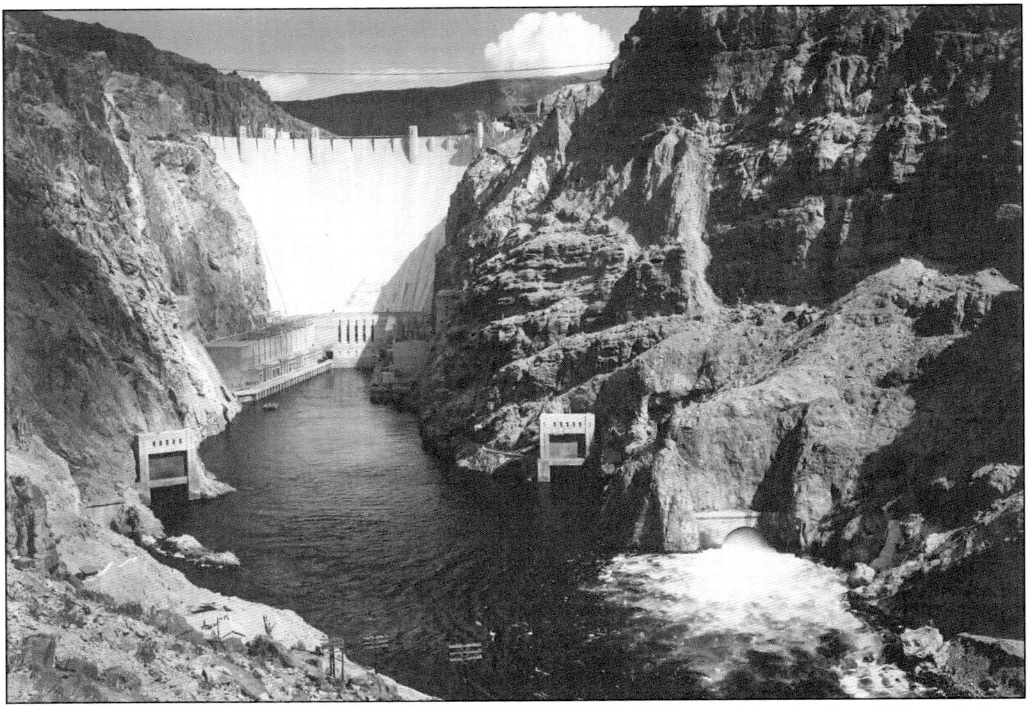

Six months after Conowingo's generators spun into service, Congress passed and Pres. Calvin Coolidge signed the Boulder Canyon Project Act, which authorized construction of what is now called the Hoover Dam, and approved a $140 million US Treasury loan to build it. (The loan was finally paid off in 1987.) The use of public funds for Hoover Dam stands in sharp contrast to earlier dams, such as Conowingo, which were built with private capital. When Hoover Dam opened on September 30, 1935, it was the world's biggest and tallest hydropower dam, rising 726 feet high and stretching 1,244 feet across the Black Canyon of the Colorado River. The marvelous setting inspired famous landscape photographer Ansel Adams to capture this stunning view of the dam in 1942. (Both, courtesy of the National Archives.)

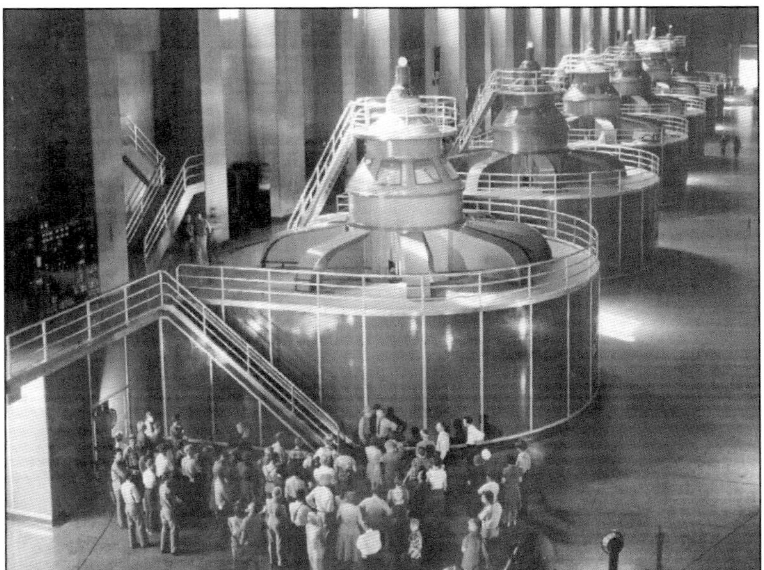

Pres. Franklin D. Roosevelt took office on March 4, 1933, in the depths of the Depression, and immediately championed big dam projects. He is pictured two months into his first term signing the act establishing the Tennessee Valley Authority (TVA). The TVA was created primarily to build hydropower plants such as Fort Loudon Dam (below) that would bring flood control and electric power to the impoverished Tennessee Valley region. It was one of Roosevelt's New Deal initiatives to use dam construction projects to get unemployed citizens back to work and lift the nation out of the Depression. Roosevelt was convinced that pump-priming, or federal spending, would stimulate the economy, and hydropower dam projects were high on his list. The TVA, the largest government-owned industrial enterprise in American history, was born. (Right, courtesy of the Library of Congress; below, courtesy of the Tennessee Valley Authority.)

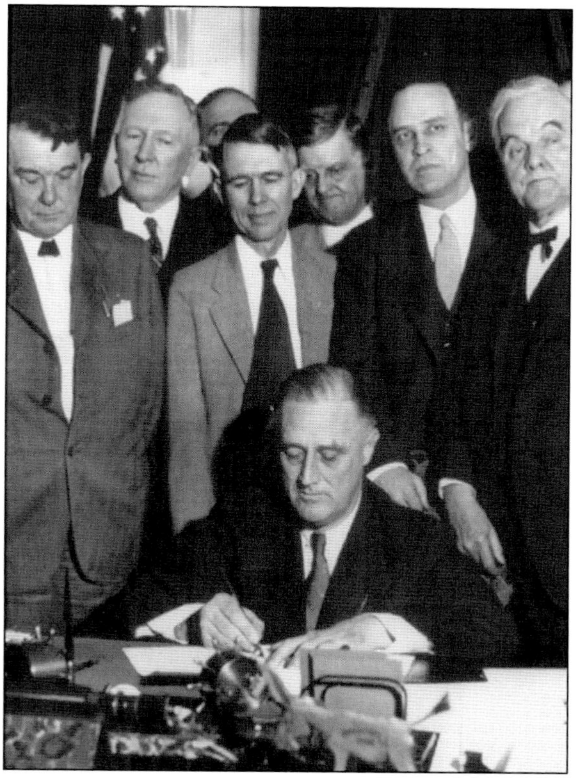

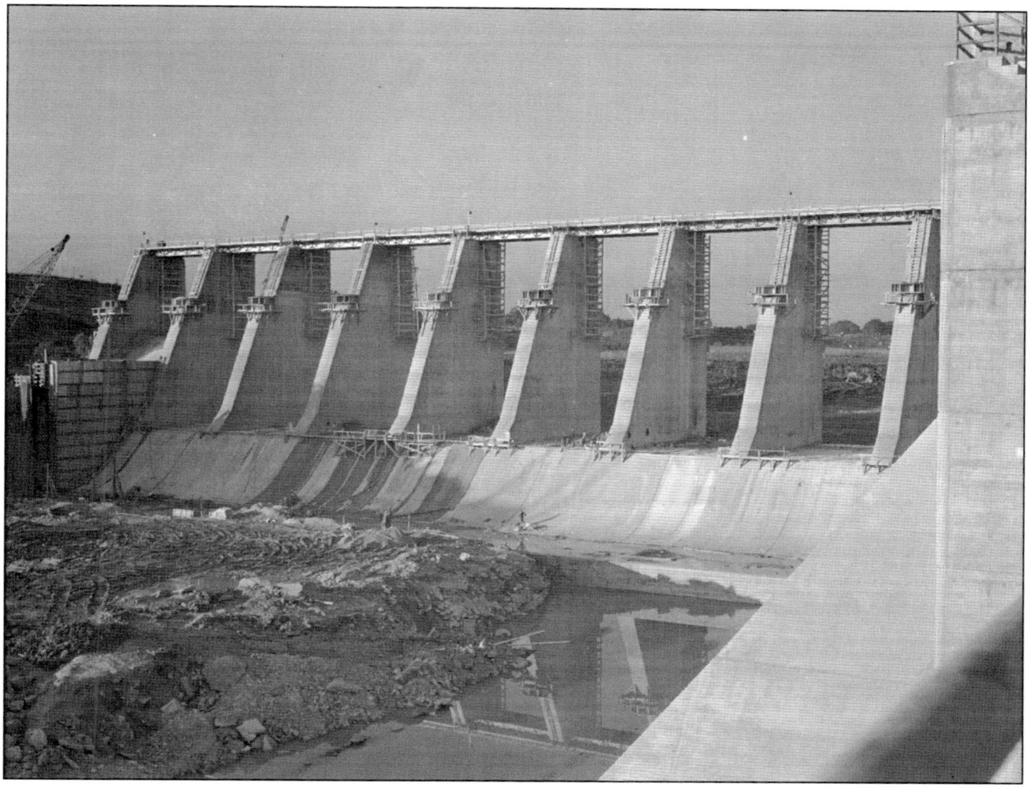

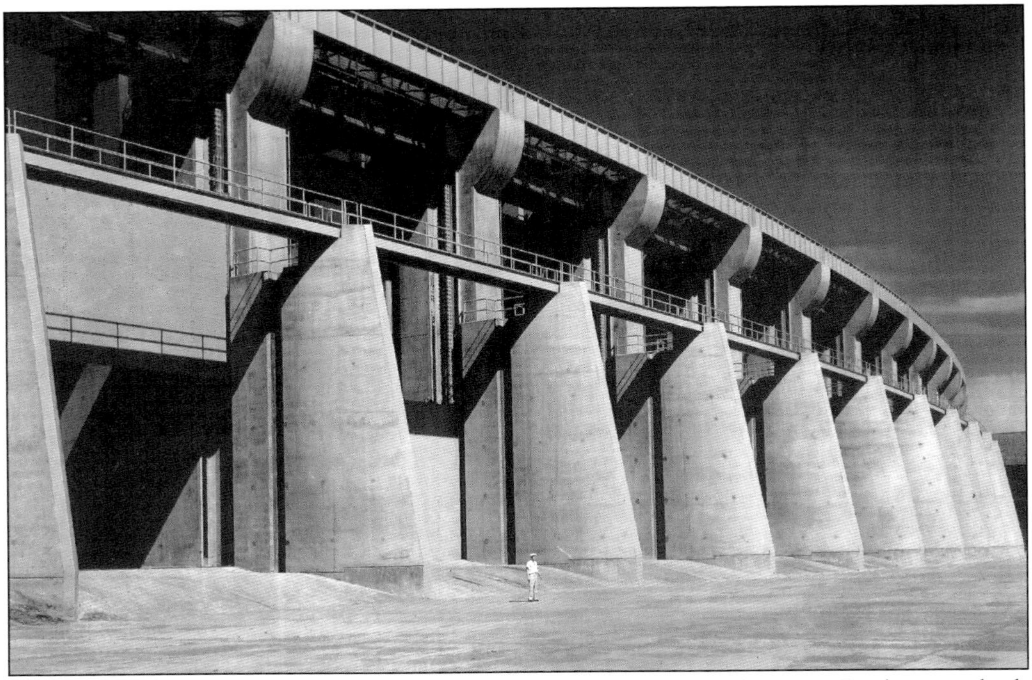

Fort Peck Dam (above) on the Missouri River in Montana was another New Deal project built under the Public Works Administration. Construction began in 1933, and President Roosevelt is seen surveying its progress in August 1934 (below). From the back of a railcar, Roosevelt declared his grand vision of hydropower as the key to America's future prosperity: "People talk about the Fort Peck Dam as the fulfillment of a dream. It is only a small percentage of the whole dream covering all of the important watersheds of the Nation. Before American men and women get through with this job, we are going to make every ounce and every gallon of water that falls from the Heaven and the hills count before it makes its way down to the Gulf of Mexico . . . . I am very confident it is going to be carried through to the success and glory of the Nation." (Both, courtesy of the US Army Corps of Engineers.)

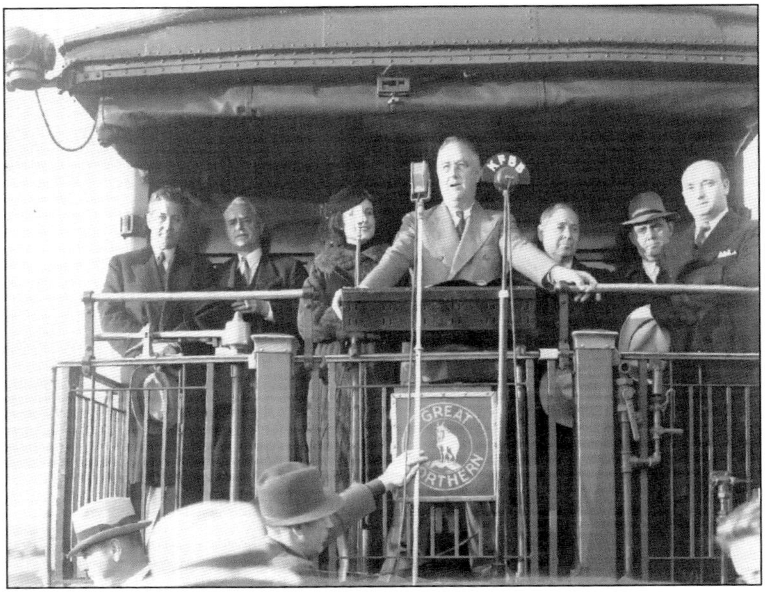

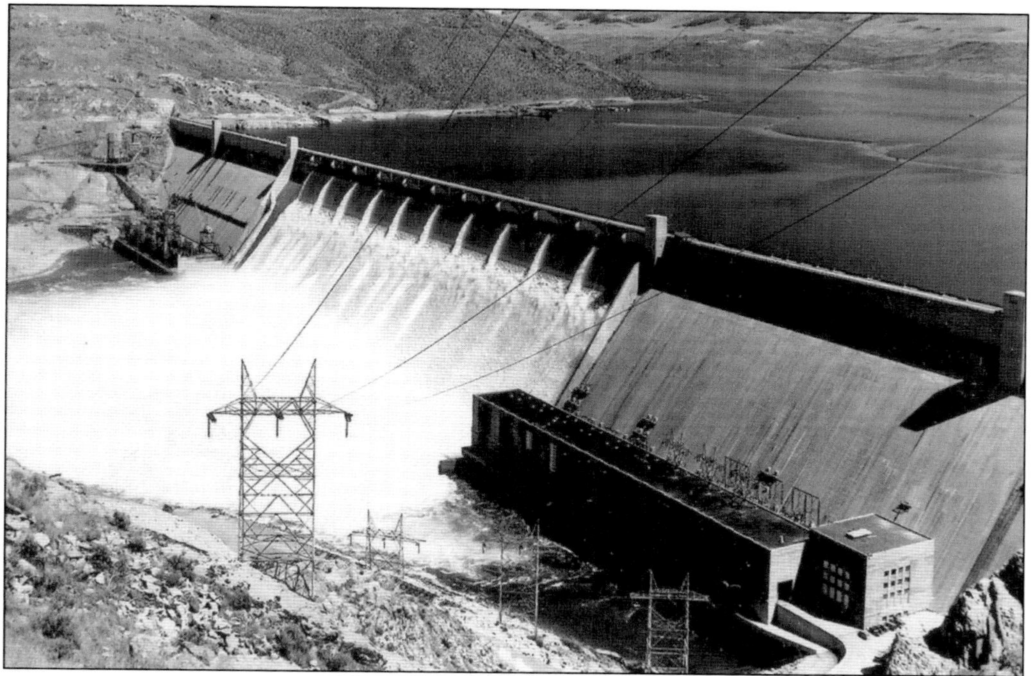

When Roosevelt ran for president in 1932, he promised that the federal government would build dams on the Columbia River to supply the Pacific Northwest with cheap electricity. After winning the presidency, he made good on his promise. The Grand Coulee Dam, seen above in 1942, was made possible by the New Deal's public works funding. Roosevelt took an especially active interest in the development of both the Grand Coulee and Bonneville Dams. The 1937 photograph below shows FDR at the Bonneville Dam. It was here, three years earlier, that he had championed the government's role in hydropower: "Over a year ago, when we first established the principle of commencing great public works projects in every part of the Union, I became firmly convinced that the Federal Government ought immediately to undertake the construction of the Bonneville Dam and the Grand Coulee Dam, and so we got started." (Above, courtesy of the US Bureau Reclamation; below, courtesy of the Bonneville Power Administration.)

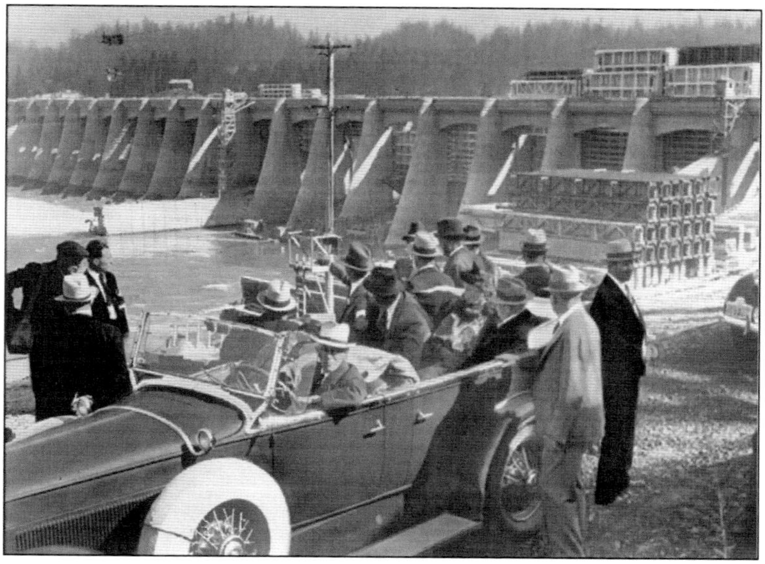

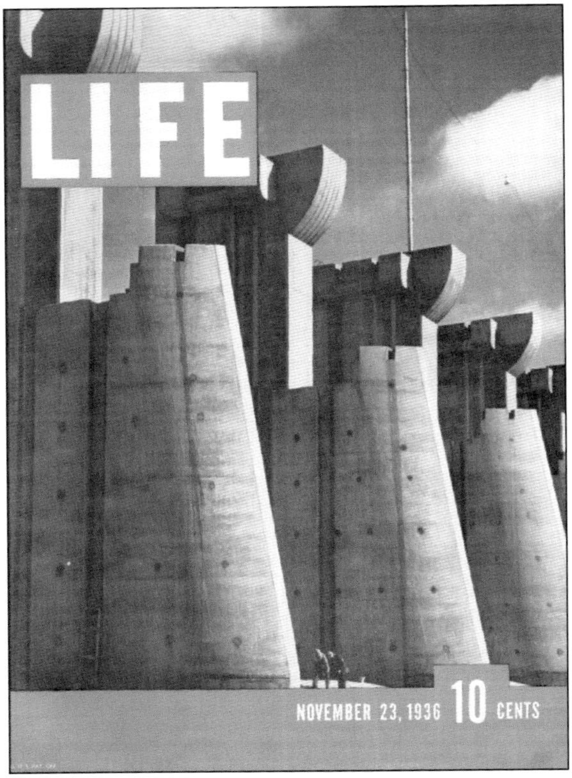

By the mid-1930s, with several prominent dam projects underway, hydropower was entering the nation's collective conscious. *Life* magazine helped fix the idea firmly in the minds of the public when its editors chose Fort Peck Dam to adorn the cover of its first issue in 1936. The celebrated photographer Margaret Bourke-White captured this view of the dam with its towering concrete abutments ascending skyward. The dam was completed and went into service in 1940. (Public domain.)

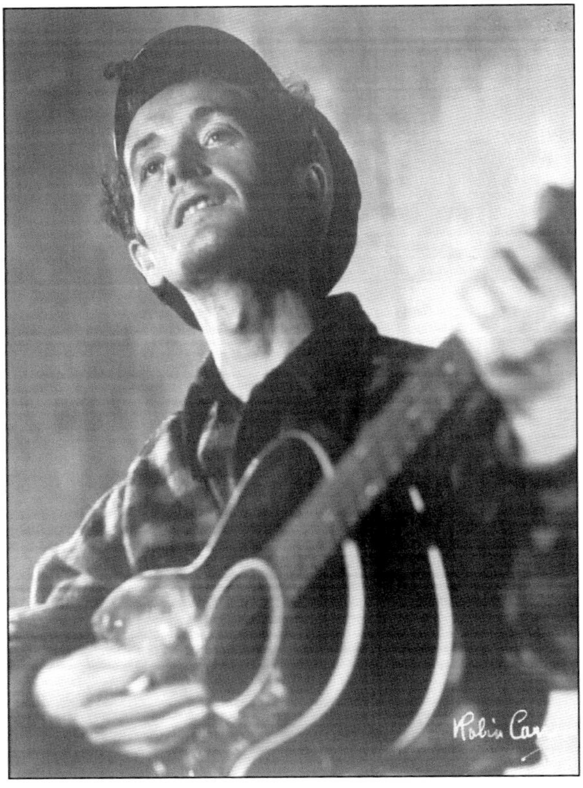

The Bonneville Power Administration promoted the dams of the Pacific Northwest with music. In May 1941, it hired folksinger Woody Guthrie to tour Grand Coulee and Bonneville Dams and write about them. Guthrie proceeded to compose songs such as "Talking Columbia," which lionizes man's harnessing of nature for power and the benefits of electrification to ordinary citizens. The following are the song's lyrics: "Uncle Sam needs houses and stuff to eat, / And Uncle Sam needs wool, Uncle Sam needs wheat, / Uncle Sam needs water and power dams, / Uncle Sam needs people and people need land. / Of course I don't like dictators not much, myself, / But then I think the whole country ought to be run / By eeeeee-lectricity!" (Courtesy of the Library of Congress.)

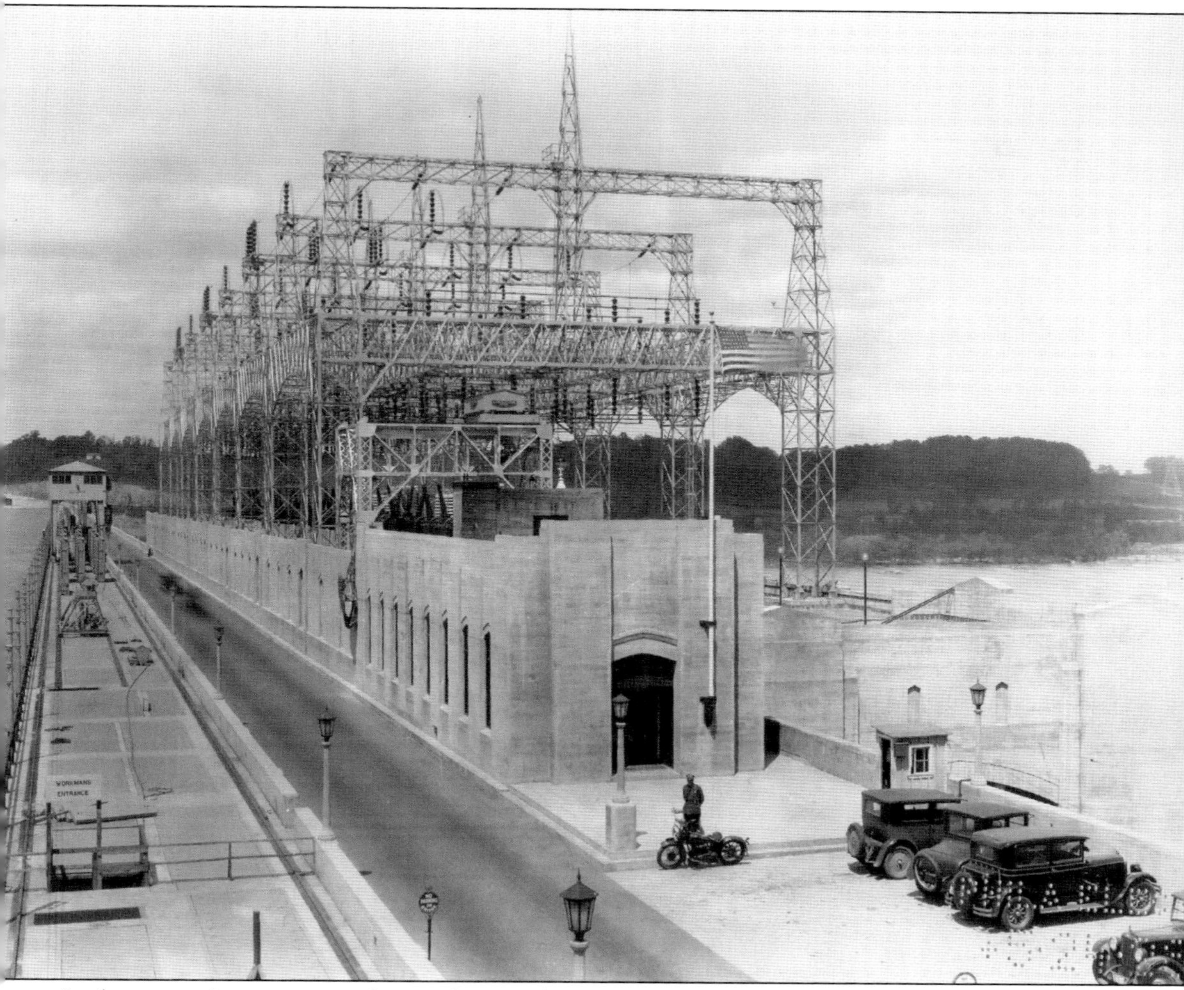

In the years after coming on line, as long as the river provided adequate flow, the hydro plant at Conowingo reliably transformed waterpower into electricity. This view of the dam was taken in 1929 from the steep bank of the Harford County side of the river and shows a patrolman with motorcycle parked at the original main entrance to the power plant. In its early days of operation, employees parked in a small parking lot here and checked in at the guard shack as they reported for duty. Prominent in this photograph, on the roof of the powerhouse, is the high-tension substation where switching facilities controlled the delivery of current from generating units to transmission lines.

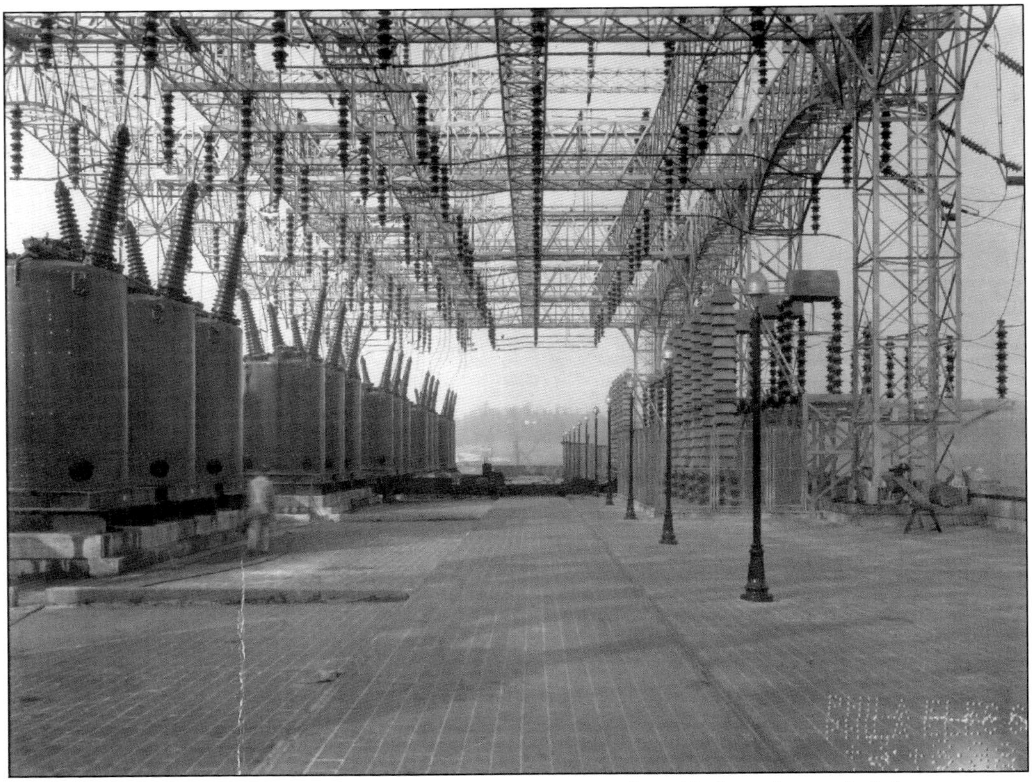

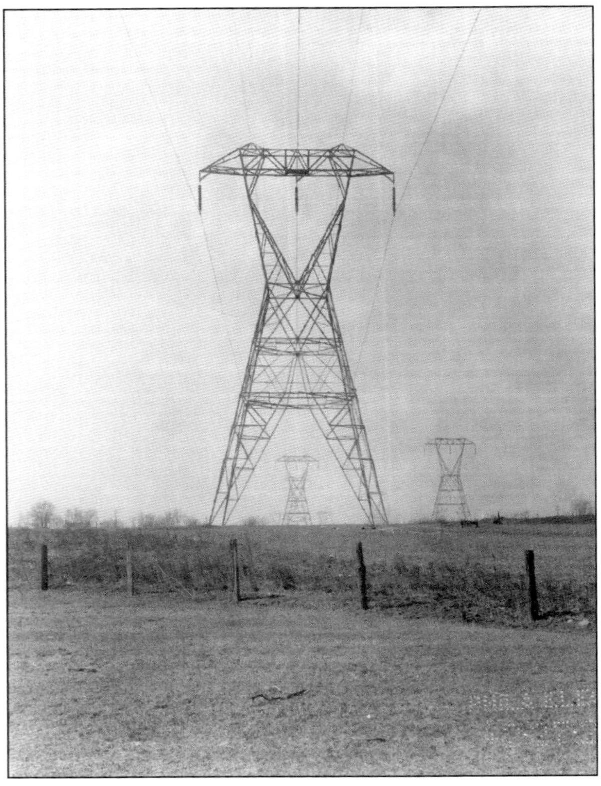

Installed on the roof of the Conowingo power plant were oil circuit breakers, transformers, and transmission lines. Generators inside turbine hall produced a three-phase alternating current that was sent up into this rooftop switching structure and then sent down power lines that stretched over the Susquehanna River to Rowland Island before turning northward toward Pennsylvania.

Conowingo's power served customers in eastern Pennsylvania and New Jersey, with the city of Philadelphia as the principal market. This was an exceptionally long power run at the time. One-inch transmission lines and galvanized steel suspension towers extended along a 58-mile right-of-way to a substation at Plymouth Meeting, Pennsylvania, about 15 miles northwest of Philadelphia, where one of the largest outdoor substations in the world was constructed.

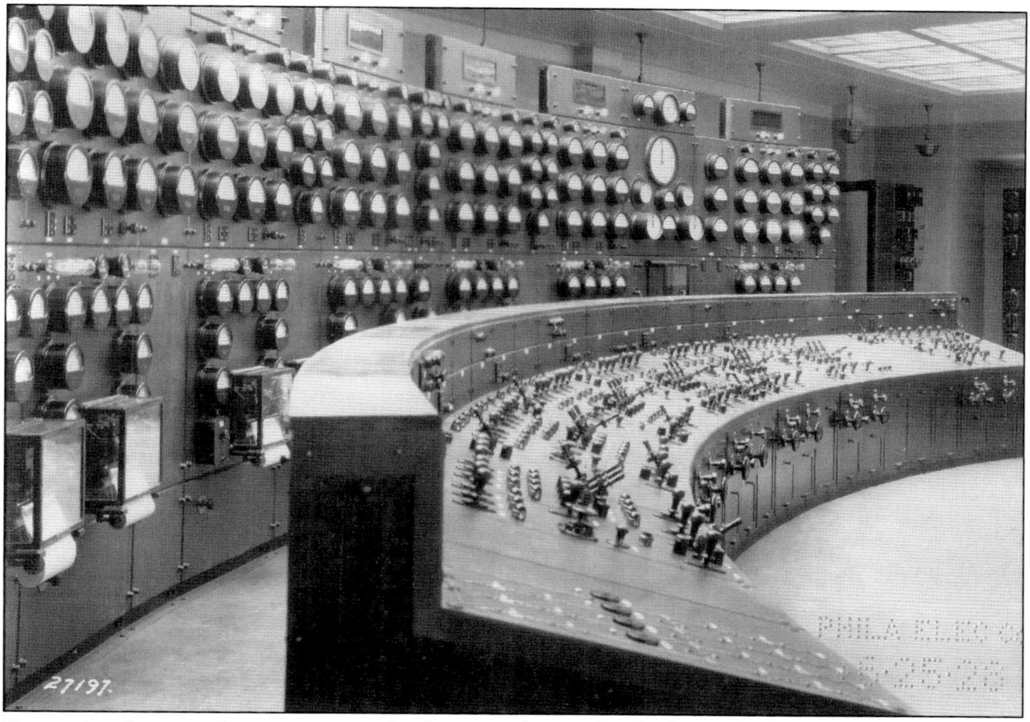

Conowingo's control room was inside the powerhouse, high above the humming generators in turbine hall. Two semicircular switchboards were installed in the room, one behind the other. As seen in the photograph above, various control switches and indicator lamps were mounted on the front bench board so that an operator could use them to start, control, and shut down the generating units. Behind the bench board was an instrument panel, pictured below, with the ammeters, voltmeters, watt meters, and reactive kva indicators necessary to monitor and control the electrical circuits. Around the rest of the room were other boards for the control of voltage-regulating equipment and gauges that measured the elevation of water in the reservoir behind the dam and the flow of water through the dam.

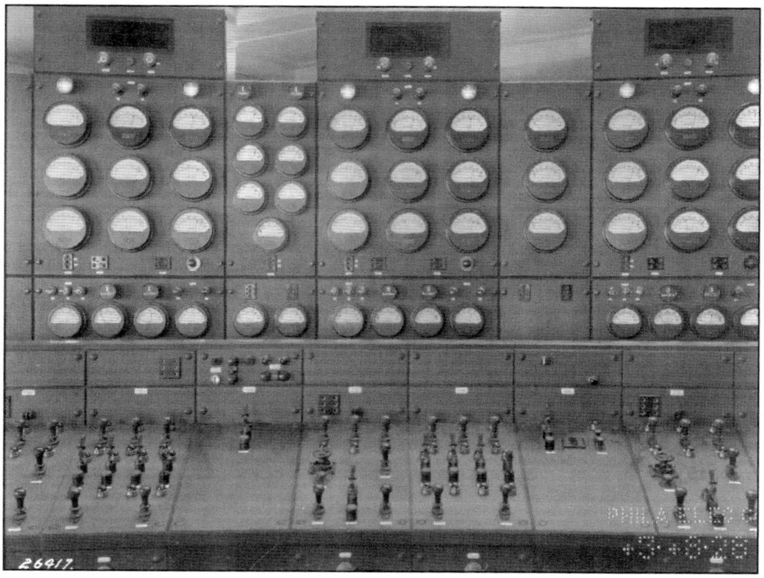

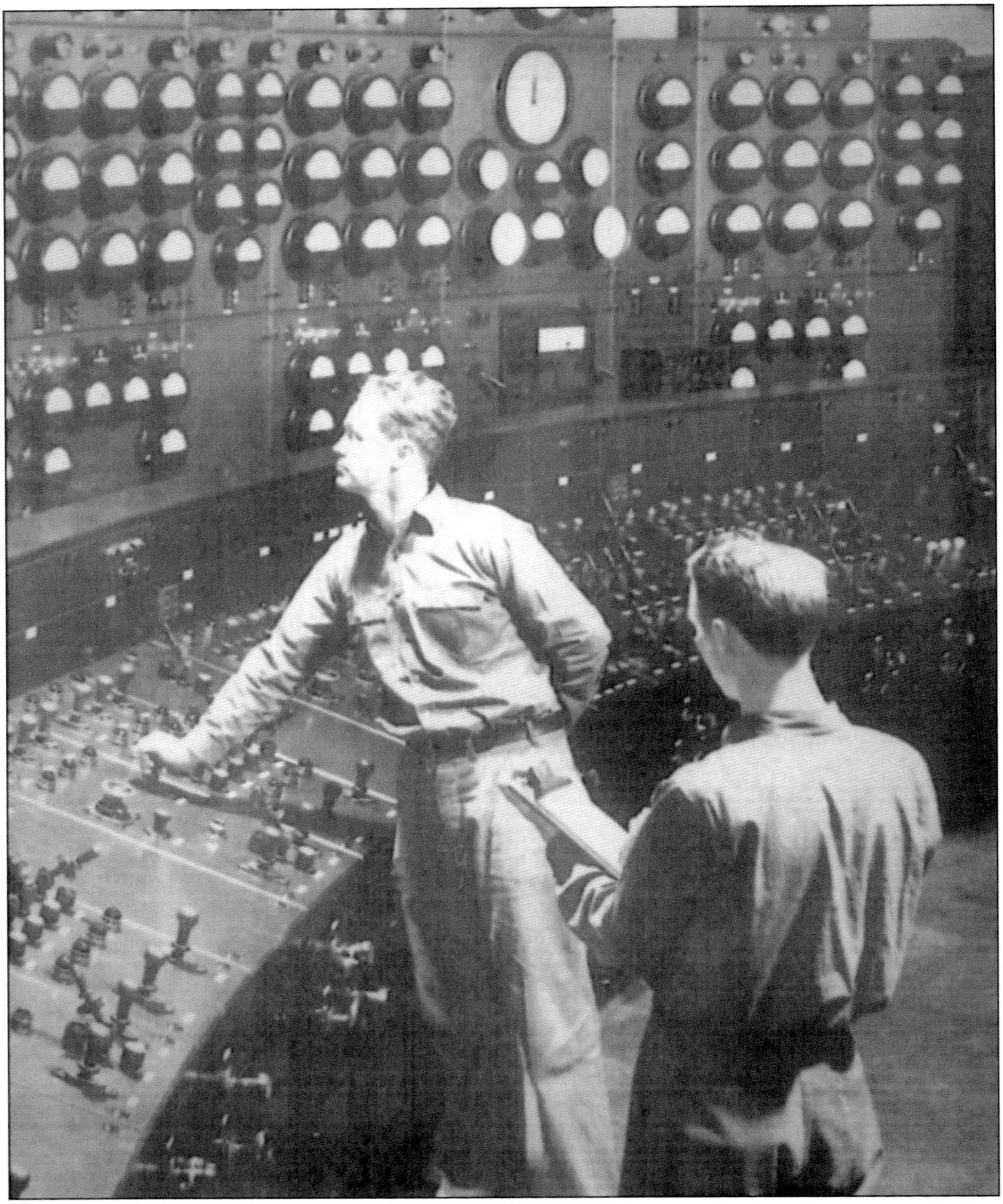

The responsibility for starting a generator, as well as the delicate task of synchronizing it before transmitting its output to the electrical grid, belonged to the control room operator, pictured here. The operation required a skilled hand and keen eye. To bring a generator online, the operator triggered a governor system to open what are called wicket gates. The wicket gates, located inside the scroll case, can be thought of as giant louvered window blinds; they open and close to fine-tune the flow of water past the turbine blades. The wider the gap of the wicket gate, the greater the flow of water and the faster the turbine spins. When the generator was running at maximum efficiency and its output frequency was "in time" with the grid, the operator brought it online. With his hand on the switch, the operator watched the needle on the voltmeter oscillate back and forth at 60 cycles per second. It was then a matter of timing; at the right moment, he threw the switch and power flowed out of the plant.

Seven

Transforming Rural Maryland

With seven generators spinning inside turbine hall and a ribbon of concrete stretching across the river, the Conowingo Dam forever changed the lower Susquehanna Valley. It brought employment to local managers, engineers, technicians, and laborers, but the man-made colossus also impeded what had always been a free-flowing river. Therefore, fish that had traditionally migrated up the river to spawn now ran into an impenetrable fortress. Sportfishing flourished as the dam's catwalks became a destination for line-and-pole enthusiasts, and an abundance of bald eagles appeared to feed upon the fish. These magnificent birds of prey became attractions themselves for throngs of bird-watchers, and tourists drove across the top of the dam to admire the miles-long view along the canyon-like walls of the Susquehanna River.

But perhaps the most indelible transformation of the region was its physical appearance. There was beauty in the way the Conowingo stretched long and low between its canyon banks and blended naturally into the landscape's rolling hills. The technological triumph of 1928 was also an aesthetic success. While it was not meant to be an architectural statement, the Conowingo Dam's designers imbued the enormous structure with certain touches that made it pleasing to behold: the symmetry of the arched windows, the curve of the spillway, the neon "Conowingo Hydroelectric Plant" glowing in the night, and the geometric shapes and angles of the transformers, steel towers, and transmission lines. Visitors purchased postcards of the dam as keepsakes, and artists and photographers were inspired to celebrate it. Every morning, the sun beamed into turbine hall, and at night, the powerhouse glowed with electricity. This new Maryland landmark, with its whirring generators and churning water, caused quite a hubbub and ushered in a new era both for the people who lived near it and those who came to work inside it.

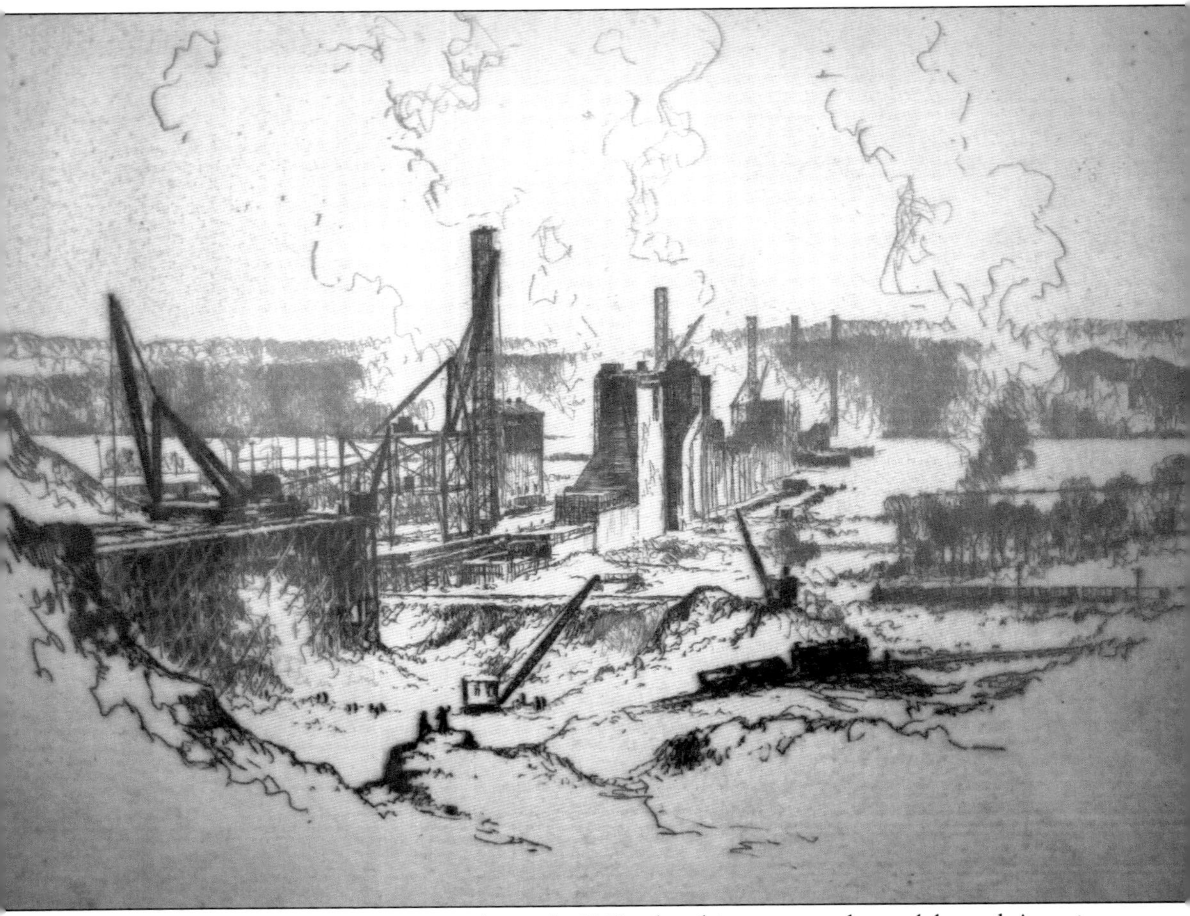

Otto Kuhler gained notoriety in the early 1920s sketching scenes that celebrated American engineering and industrial might. He saw beauty in the swirling clouds of progress at Conowingo Dam and expressed it artfully in this plate etching, *Harnessing the Susquehanna at Conowingo, Maryland*. The etching was printed by D.C. Elphinstone, Inc., of Baltimore, Maryland, who wrote, "These water power developments are getting bigger and bigger. It's a matter of more and more power – and of dollars and cents. We may forget the romance of it. But there are men who do not. 'The working world' is O. Kuhler's territory. He came to America six years ago and, starting in Pittsburgh, began drawing 'America at work.' I asked him to do one etching for me. He did one plate, of which this is the 69th impression of an edition of 75, and then destroyed the plate."

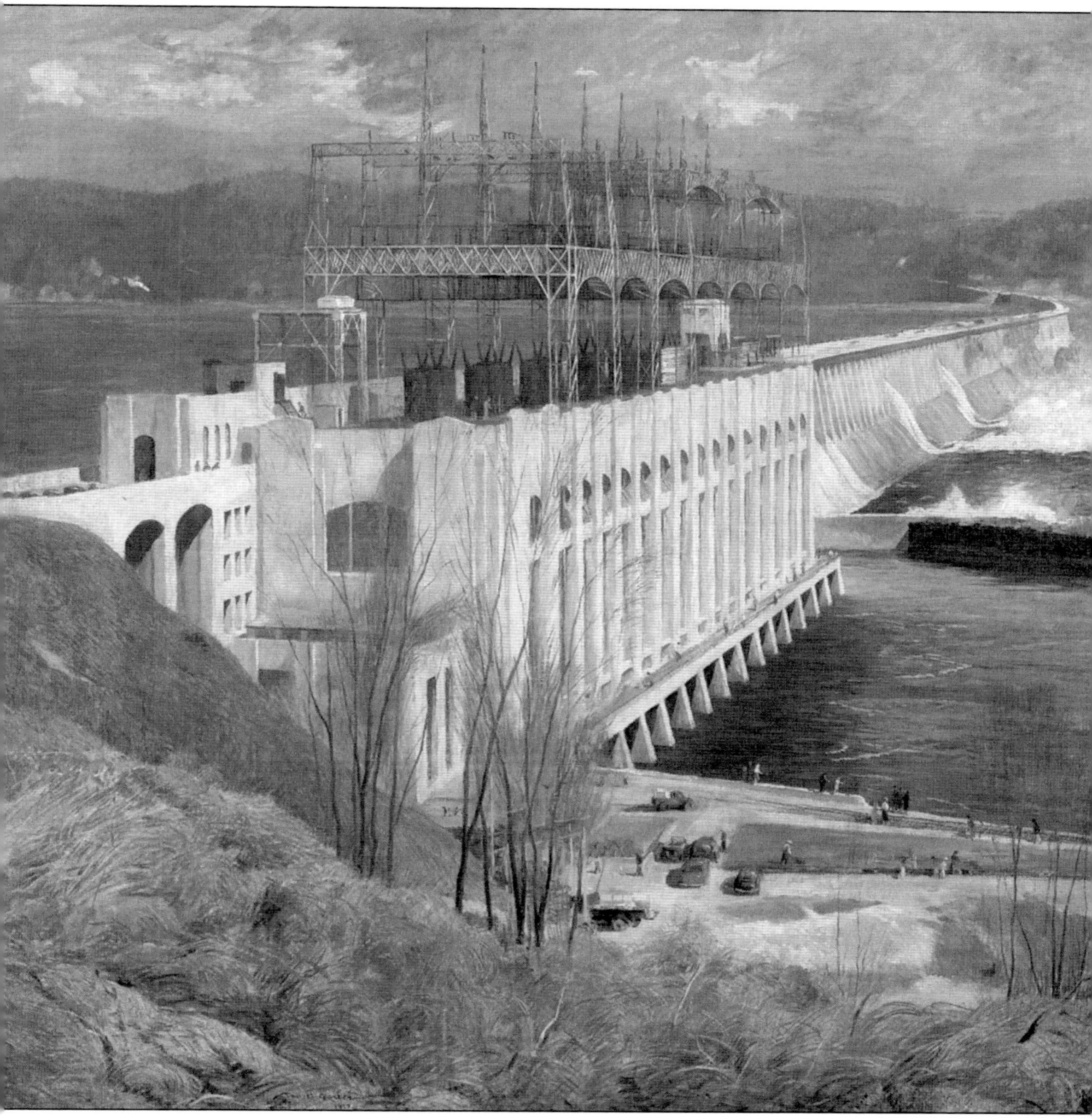

There is perhaps no more romantic a rendering of the dam than that by landscape painter Daniel Garber. Garber was an American Impressionist once lauded as "a Modern American Master." An analysis of "the law of Daniel Garber's mind" by *Art & Life* magazine claims he had "discovered in himself the ardent desire for pictorial expression [that] will give sometimes his own interpretation of natural objects with almost childlike clearness of perception." The magazine continues with Daniel Garber's artistic credo: "An artist must try to share with his fellow-creatures the happiness he has received from the study and interpretation of nature." Garber's application of that aesthetic to the Conowingo Dam, painted when it was just completed, stands as a reward for all Americans.

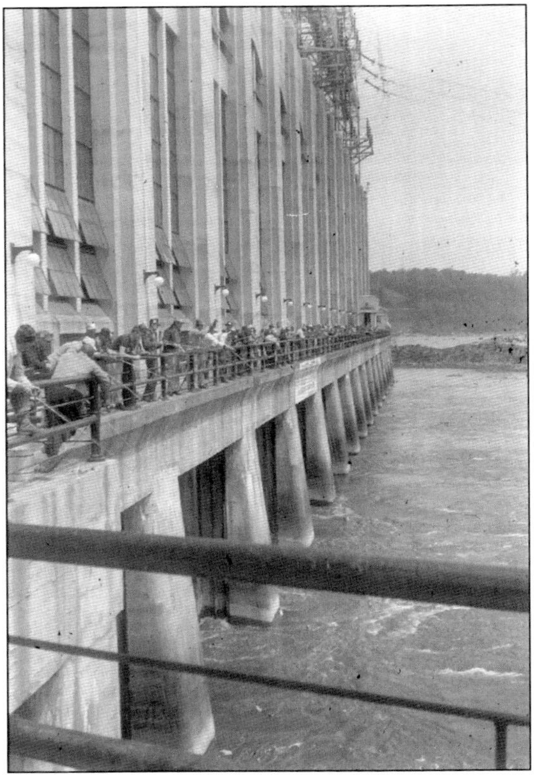

Fishermen were immediate beneficiaries of the dam, as indicated by these photographs from 1956. Conowingo Pond was widely praised for its good fishing, and in 1930, the pond was stocked with 20,000,000 yellow perch to make it even better. But the best fishing was directly below the dam, where fishermen were once allowed to hike onto the catwalk and drop their lines into the water that spewed from tailrace. It was an especially productive spot, for when a generator spun into motion, more than just water passed through it. Small bait fish were sucked down from the pond above the dam, and after a surreal trip through the turbines, they would emerge stunned. Catfish, bluefish, and striped bass lurking in deep pools below the dam would then move up to feed on the dazed fish, and in the ensuing feeding frenzy, anglers lowered their lures and had a very good chance of landing a lunker.

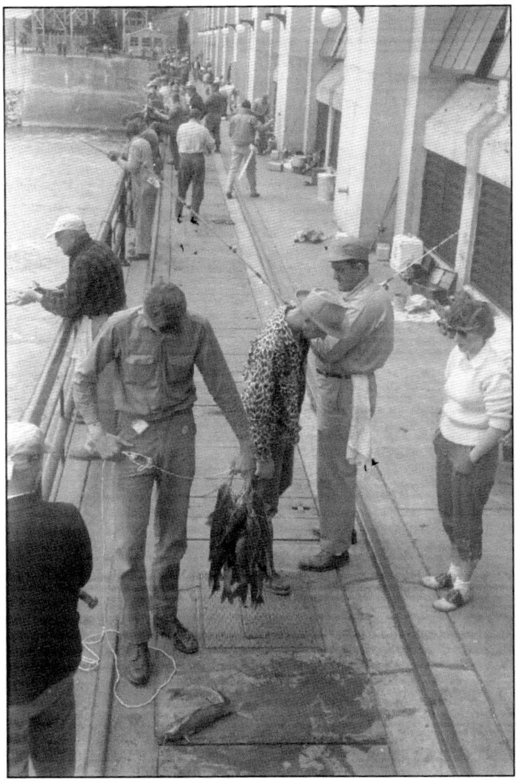

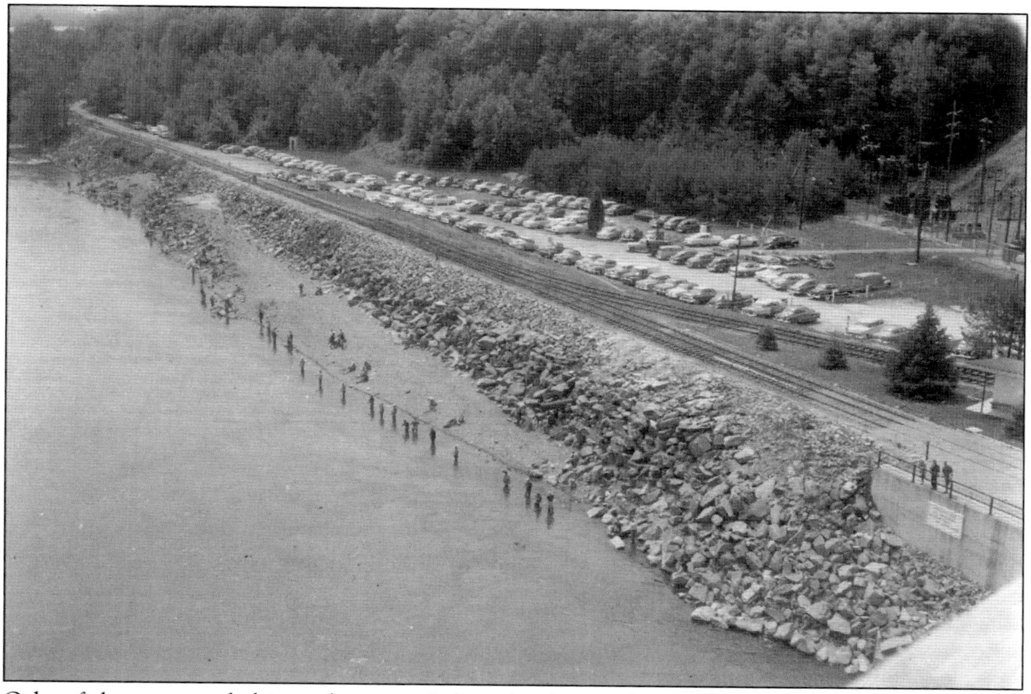

Other fishermen waded into the water below the dam and cast their bait up into the tailrace. It was common for a multitude of people in hip-waders to be lined up in the water below the dam, casting heavy lures back at it. They shared the limited territory with equanimity, perhaps because they knew they would all be equally rewarded by an impressive catch. However, these anglers had to be aware that conditions at the tailrace could rapidly turn treacherous, for when the plant fired up a generator, the water level below the dam would come up dangerously fast. A system was developed at Conowingo to warn the waders; when the red lights flashed and sirens blared, beware. One moment, the river would be peaceful, and then water would gush suddenly and violently out the tailrace into a roiling, surging current that could carry away anyone who failed to heed and retreat.

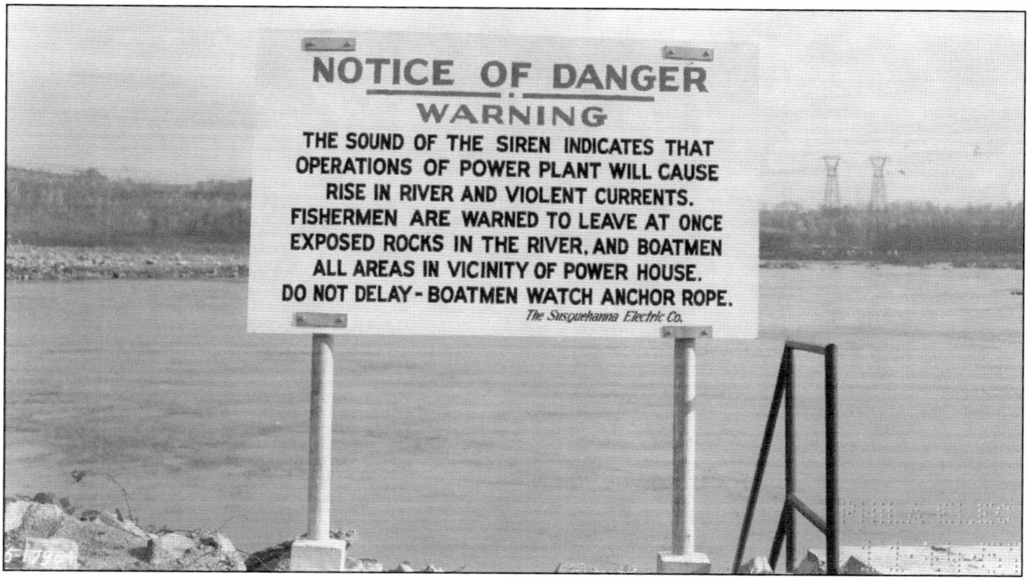

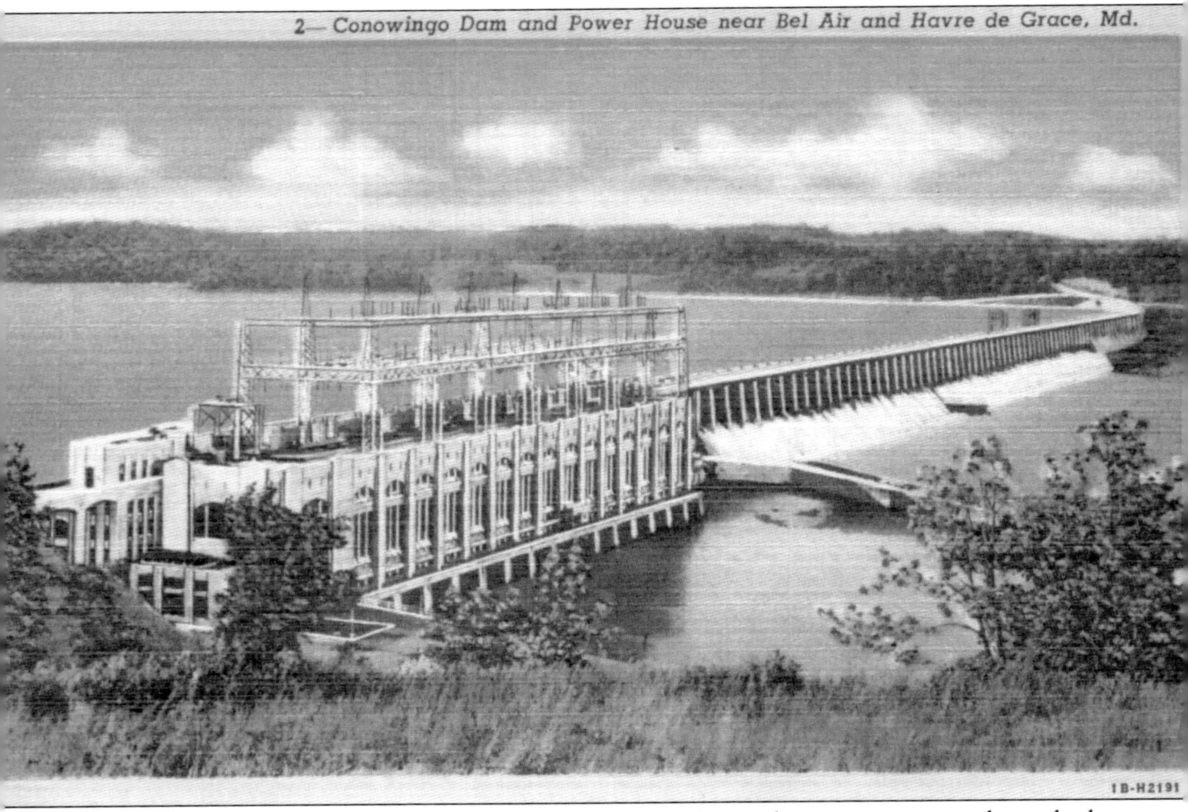

In the early 20th century, picture postcards became wildly popular among tourists who wished to share their vacations with loved ones and to report the sights they had seen along the way. Travel postcards with bucolic views, such as this one of Conowingo Dam, were manufactured to promote every imaginable destination. The postmaster general made it economical to mail postcards; a visitor to the newly opened Conowingo Dam in the late 1920s and 1930s could send this picture postcard to family and friends for 1¢. People also collected the postcards as souvenirs, to remind them of the memorable landmarks they encountered during their journeys.

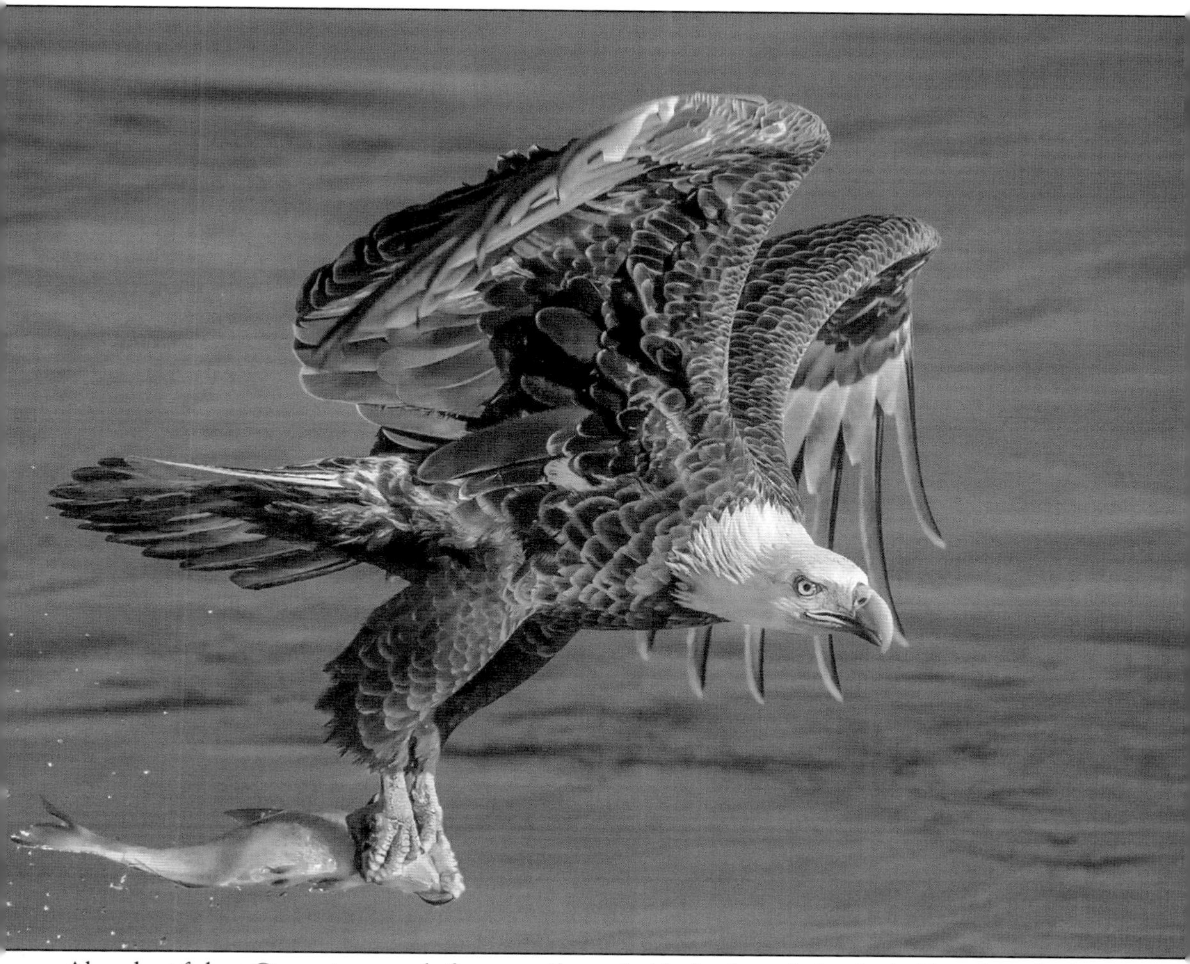

Abundant fish at Conowingo made for an easily catchable food supply for the bald eagles, which began nesting in significant numbers in the nearby forest. They are like most raptors in that they prefer not to work very hard for their meals. But when the turbines started spinning and stunned baitfish and their predators appeared, the eagles swept down from their nests to snatch fish from the water. Bird-watchers and photographers flocked to the dam, and it became a premier location for sighting and photographing America's national bird. This photograph, taken at Conowingo and titled *Grab and Run*, shows a bald eagle moments after making the catch, with its outstretched talons clutching its next meal. (Courtesy of Jerry am Ende.)

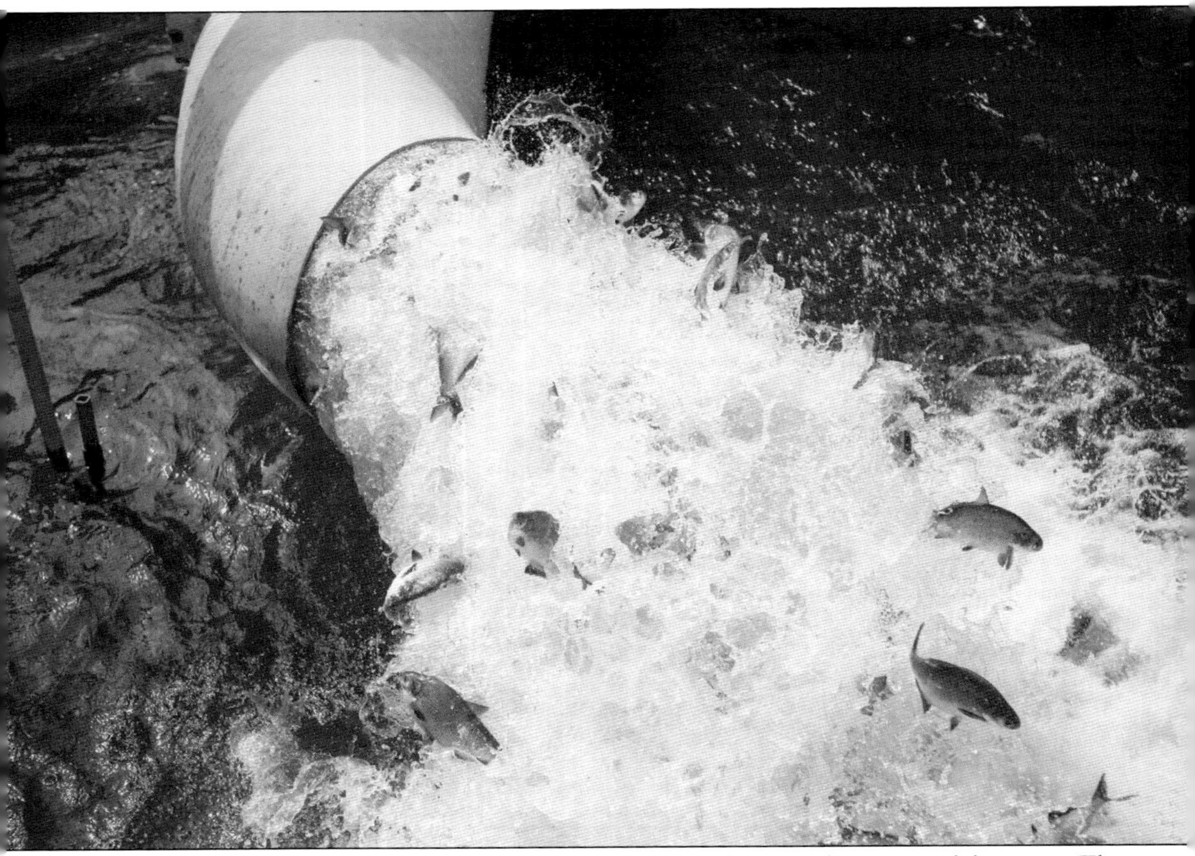

When the nearly 100-foot-high dam was built, it definitively stopped migratory fish species. The massive runs of American shad, which spawned as far up the Susquehanna River as Binghamton, New York, over 300 miles away, were suddenly over. In the 1950s and 1960s, various methods were studied to restore the shad migration, either by trucking the shad from the Chesapeake Bay to the upper Susquehanna River, or by installing fish passages over the dam. In 1968, the first fish trap was installed catching fish at the bottom of the dam and hauling them in a giant bucket to the top for release. In the mid-1970s, Conowingo installed a more efficient lift system. The fish swam into a hopper and rode to the top of the dam, where their migratory journey upriver continued. Each year, the fish are counted; and during the 2001 migratory season, 193,574 American shad passed through the Conowingo fish lift.

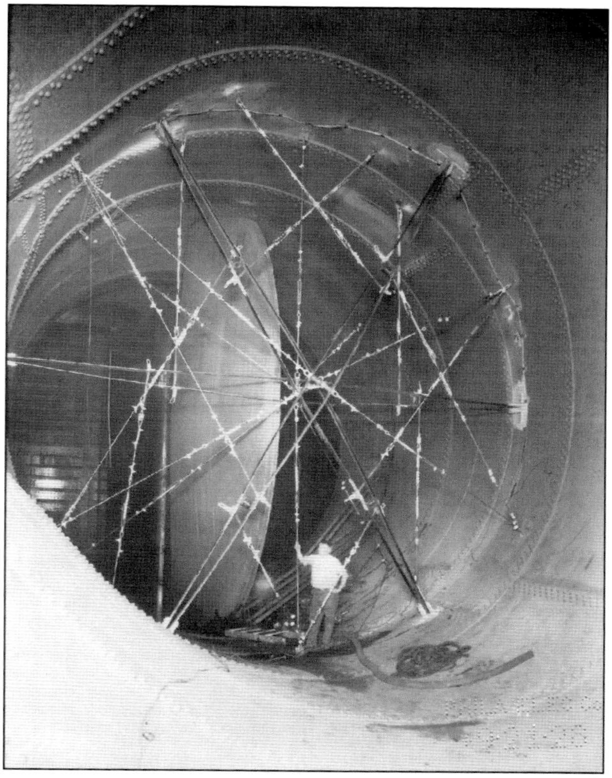

Time and the elements took an inevitable toll on Conowingo Dam. Maintaining, repairing, and replacing its massive equipment were routine aspects of the dam's operations. In this photograph, repair work is being done on the inside of the scroll case. Just beyond the giant butterfly valve are the head gates, which weigh 60 tons and hold back the immense pressure of the river.

In this photograph from the 1970s, a welder works to rehabilitate a waterwheel. When running, the force of water flowing against the runner blades causes enough stress to eventually wear the blades down. The runners were periodically removed from the scroll case for the purpose of repairing fatigue fractures and the effects of cavitations due to such stress.

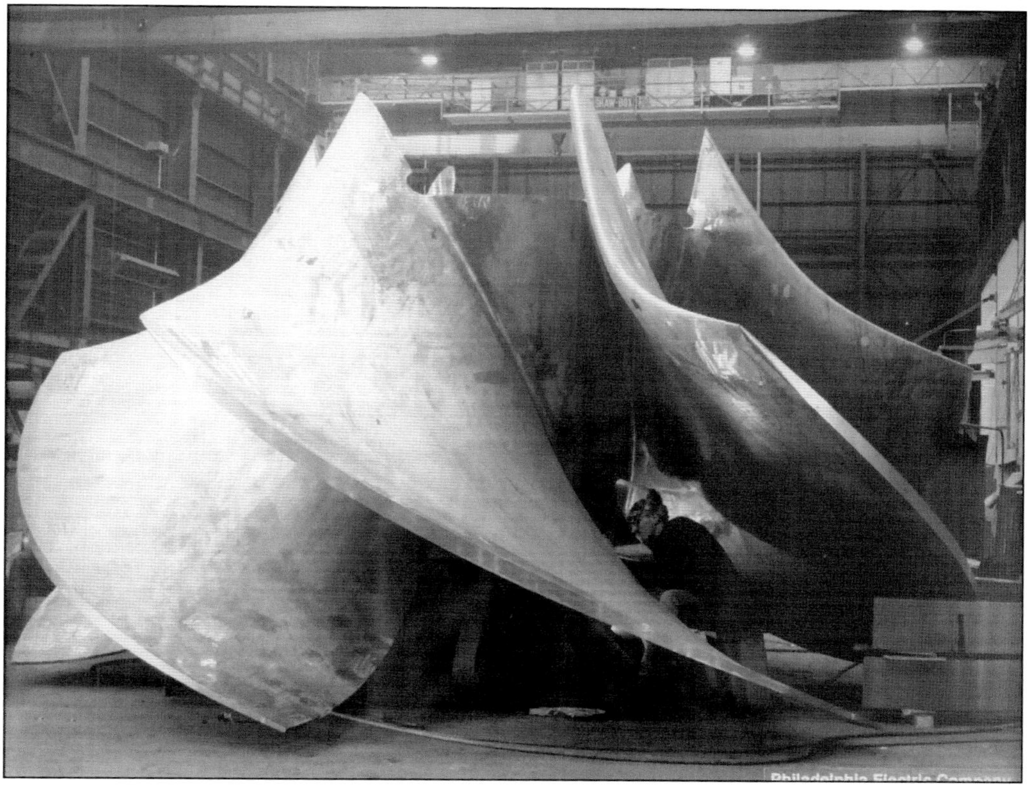

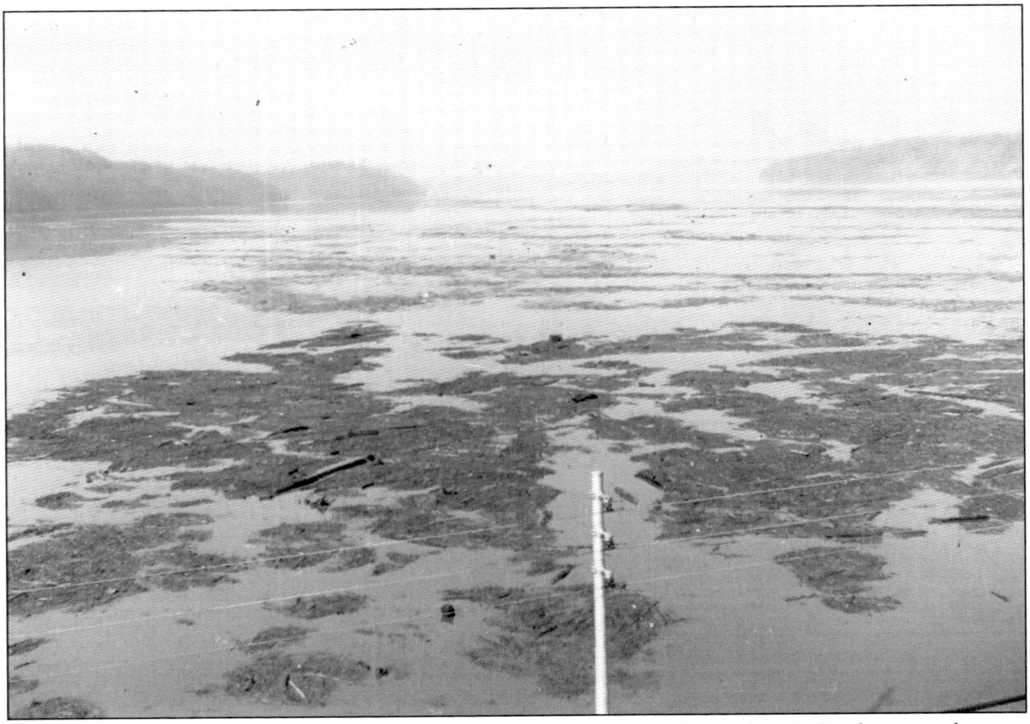

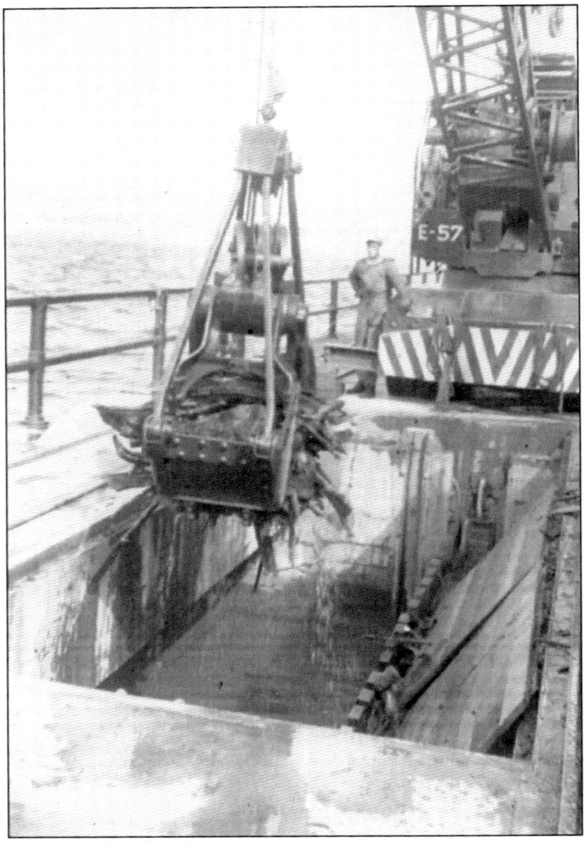

This November 1950 photograph shows the debris that inevitably floated downriver and accumulated in the forebay of the dam's headworks. If such trash were sucked through the intakes, the turbine blades could be damaged, so the detritus had to be periodically removed.

Trash racks below pond level and in front of each unit intake were designed to catch underwater debris before it entered the penstock. Trash that accumulated in front of the trash racks was cleared by means of cranes, as seen in this photograph. In this way, water could flow unimpeded through the generators.

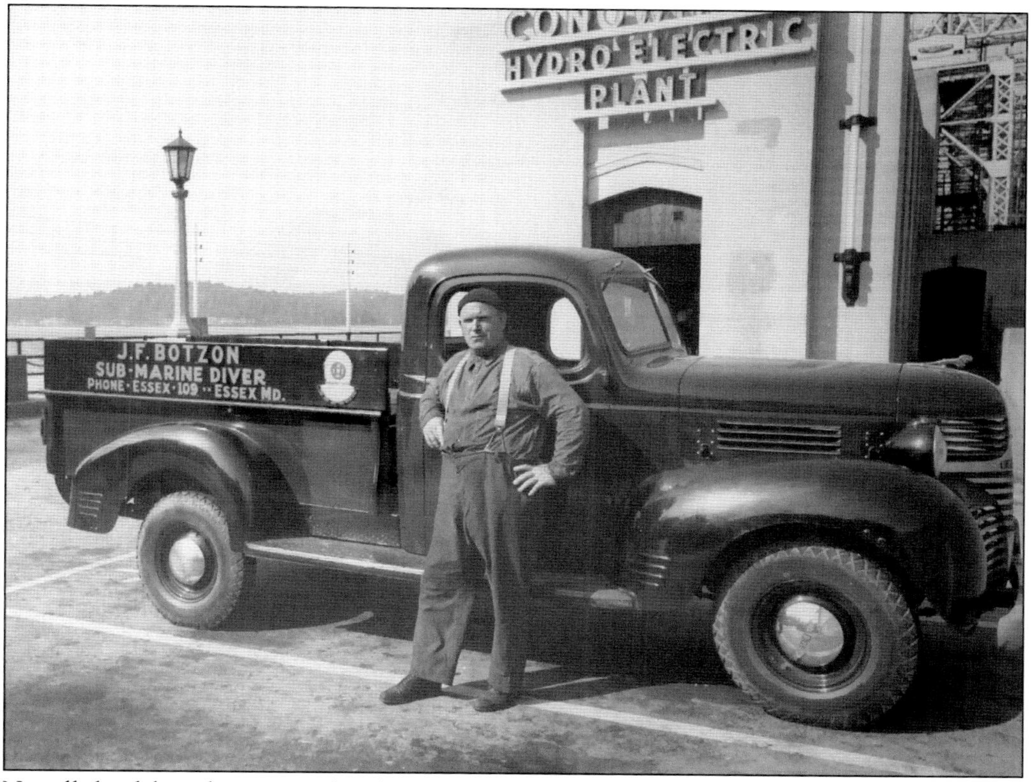

Not all the debris that accumulated at Conowingo could be cleared by cranes. Divers entered the forebay to depths nearing 100 feet, breathing air through supply lines connected to the surface, and cut the debris loose from the trash racks using chain saws. Visibility at such depths was about one foot, leading diver J.F. Botzon to say, "It's like sticking your head into a bottle of ink when you go down."

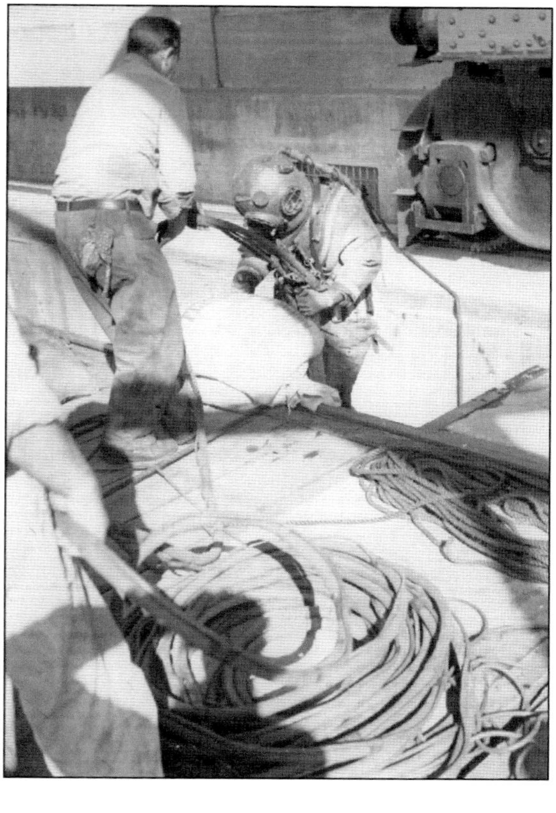

This was dangerous work. One man performing a dive in the 1970s spotted debris lodged in a gate and radioed to ask that the gate be opened. However, there was no water inside the unit, and with the immense pressure differential, the man was sucked through an 11-inch opening and killed.

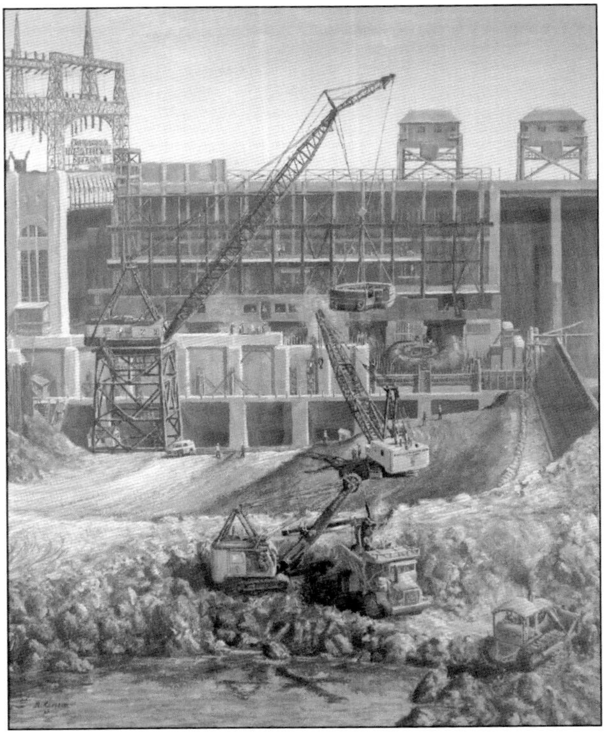

In the years since Conowingo came online, electric energy consumption has skyrocketed. During the 1940s and 1950s, Conowingo Dam's output was sufficient, but during the 1960s, the unit cost of power to consumers fell and the demand for energy surged. To keep up with the demand, four higher-capacity turbines were installed in the Conowingo plant in 1964, increasing its total output to 512 megawatts. The Philadelphia Electric Company had planned for these additional generating units in the dam's original design, thus the space to install them already existed at the eastern end of the powerhouse. New cofferdams were constructed to "unwater" this section of the dam, and during a period of three years—longer than it took to build the original dam—the new units were installed and placed into service.

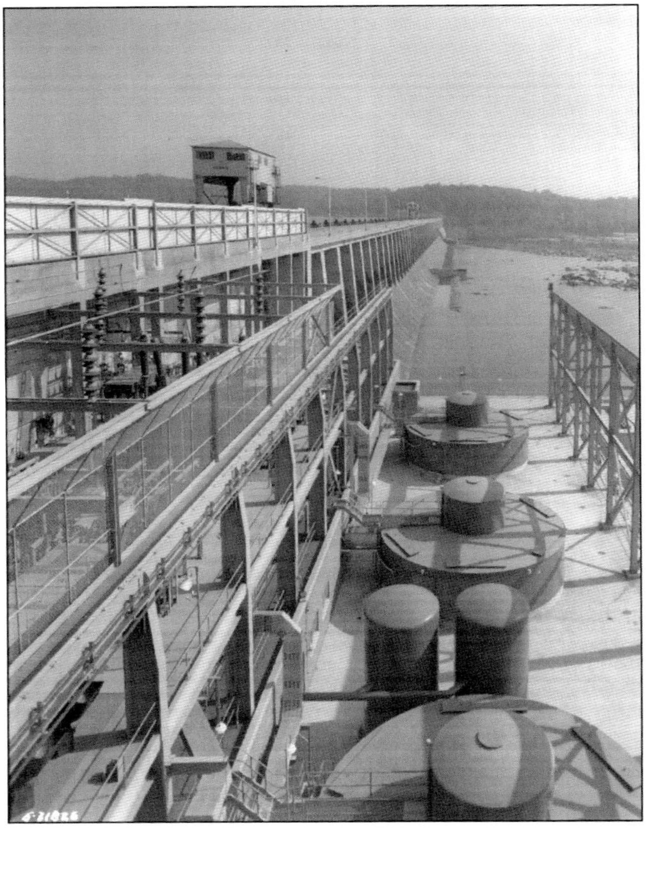

Eight

Hurricane Agnes

From the moment the Conowingo Dam roared to life in the 1920s and continuing through the Great Depression, World War II, and the 1950s and 1960s, the Susquehanna River flowed, the turbines spun, and reliable energy was created from a renewable resource. Inside the control room, the dam operator managed the plant's myriad mechanical functions, but one thing the operator could not control was the amount of water Mother Nature sent down to the dam.

At a hydro plant, elevation is an oft-used term measuring the height in feet from the bottom of the dam. For example, the base of Conowingo is elevation zero, the waterwheels are elevation 24, the floor of the powerhouse is elevation 46, and the very top of the spillway is elevation 115. The operator kept Conowingo Pond behind the dam at elevation 108.5 for maximum efficiency, and maintaining that elevation during a high-water surge was critical. Water coming downriver was first sent through the turbines, and with all its generators running, the plant passed water at a maximum rate of 80,000 cubic feet per second. If flow was greater than that, crest gates on the spillway had to be opened to prevent the pond from rising and to alleviate the pressure building against the dam.

The 50 crest gates were rated to have a combined discharge capacity of 800,000 cubic feet per second at elevation 108.5. In the historical data available to Conowingo's builders, the highest known flow of the Susquehanna River was 750,000 cubic feet per second, recorded in 1889 during the same storm that caused the devastating Johnstown Flood. Engineers therefore made Conowingo a mighty fortress designed to hold back "a flood greater than had occurred on the river in the last one hundred years." But they could not have envisioned the freak summer storm of 1972, which sent more water down the Susquehanna watershed than at any other time in the dam's history.

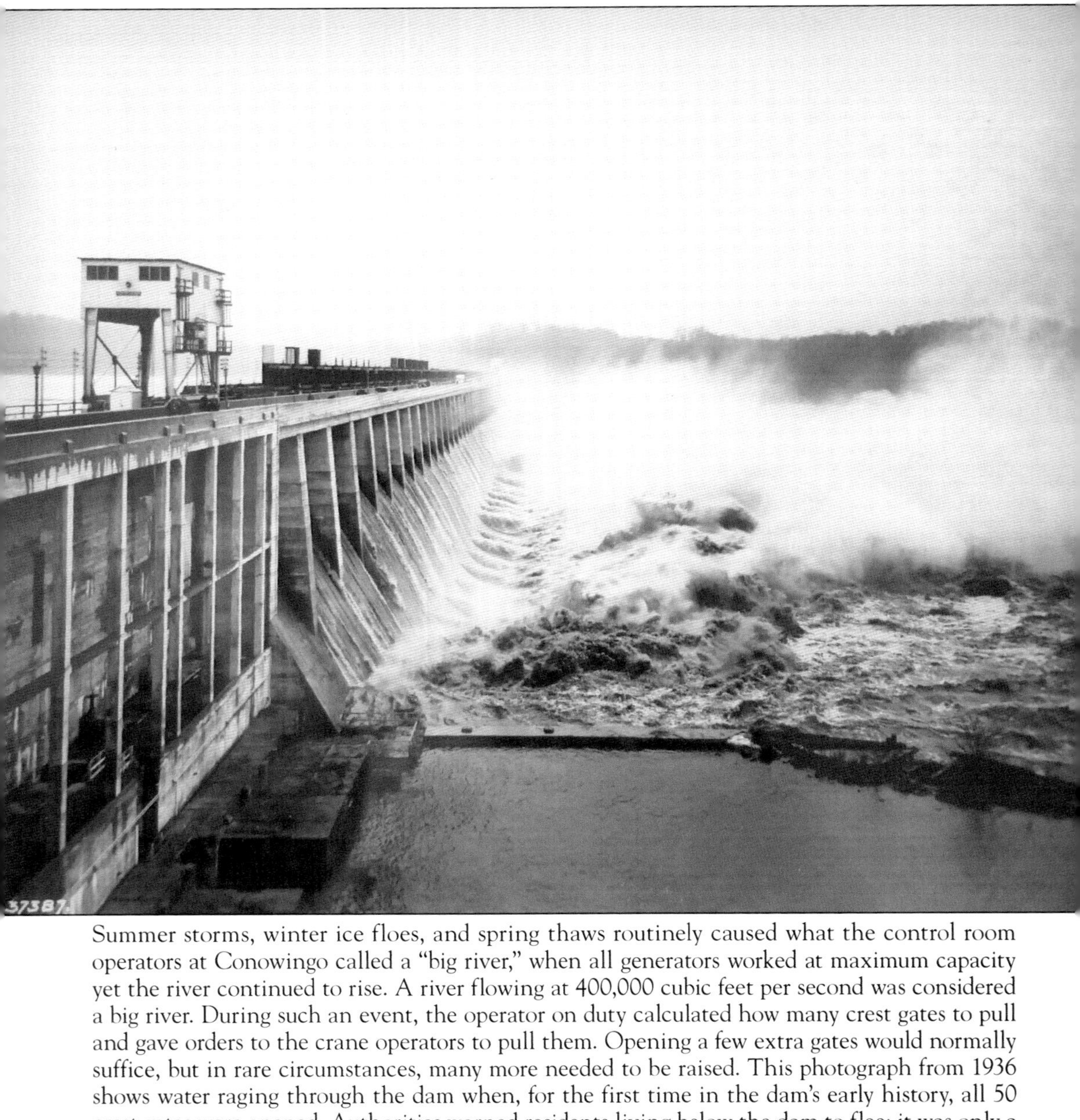

Summer storms, winter ice floes, and spring thaws routinely caused what the control room operators at Conowingo called a "big river," when all generators worked at maximum capacity yet the river continued to rise. A river flowing at 400,000 cubic feet per second was considered a big river. During such an event, the operator on duty calculated how many crest gates to pull and gave orders to the crane operators to pull them. Opening a few extra gates would normally suffice, but in rare circumstances, many more needed to be raised. This photograph from 1936 shows water raging through the dam when, for the first time in the dam's early history, all 50 crest gates were opened. Authorities warned residents living below the dam to flee; it was only a matter of time before their towns would be flooded.

All 50 gates were opened again when the Conowingo faced Hurricane Agnes, Maryland's "storm of the century." The satellite image is of the hurricane as it hit the Florida panhandle on June 19, 1972. Agnes then rolled up the Eastern Seaboard, weakened into a tropical storm, and pushed off the Atlantic coast. Then, the unexpected happened. Agnes intensified again, turned back inland, and parked over Pennsylvania and New York State. The ensuing four-day-long deluge over the Susquehanna watershed posed the worst threat in Conowingo's 44-year history. Tributaries engorged by six, twelve, and up to nineteen inches of rainwater forced the Susquehanna to overflow its banks, and as it roared down the watershed, the river narrowed to a dangerous choke point created by the Conowingo Dam. (Right, courtesy of the *Baltimore Sun*; below, courtesy of the National Oceanic and Atmospheric Administration.)

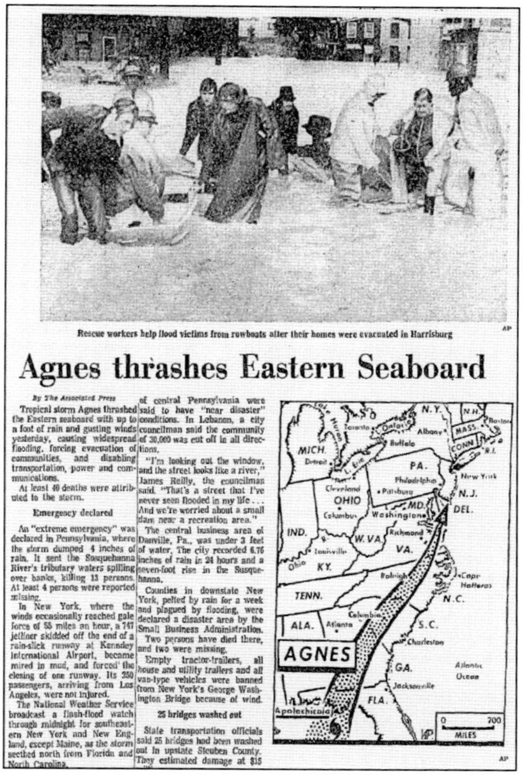

Rescue workers help flood victims from rowboats after their homes were evacuated in Harrisburg AP

Agnes thrashes Eastern Seaboard

By The Associated Press

Tropical storm Agnes thrashed the Eastern seaboard with up to a foot of rain and gusting winds yesterday, causing widespread flooding, forcing evacuation of communities, and disabling transportation, power and communications.

At least 40 deaths were attributed to the storm.

Emergency declared

An "extreme emergency" was declared in Pennsylvania, where the storm dumped 4 inches of rain. It sent the Susquehanna River's tributary waters spilling over banks, killing 13 persons. At least 4 persons were reported missing.

In New York, where the winds occasionally reached gale force of 55 miles an hour, a 747 jetliner skidded off the end of a rain-slick runway at Kenndey International Airport, became mired in mud, and forced the closing of one runway. Its 350 passengers, arriving from Los Angeles, were not injured.

The National Weather Service broadcast a flash-flood watch through midnight for southeastern New York and New England, except Maine, as the storm seethed north from Florida and North Carolina.

... of central Pennsylvania were said to have "near disaster" conditions. In Lebanon, a city councilman said the community of 30,000 was cut off in all directions.

"I'm looking out the window, and the street looks like a river," James Reilly, the councilman said. "That's a street that I've never seen flooded in my life . . . And we're worried about a small dam near a recreation area."

The central business area of Danville, Pa., was under 3 feet of water. The city recorded 6.76 inches of rain in 24 hours and a seven-foot rise in the Susquehanna.

Counties in downstate New York, pelted by rain for a week and plagued by flooding, were declared a disaster area by the Small Business Administration.

Two persons have died there, and two were missing.

Empty tractor-trailers, all house and utility trailers and all van-type vehicles were banned from New York's George Washington Bridge because of wind.

25 bridges washed out

State transportation officials said 25 bridges had been washed out in upstate Steuben County. They estimated damage at $15

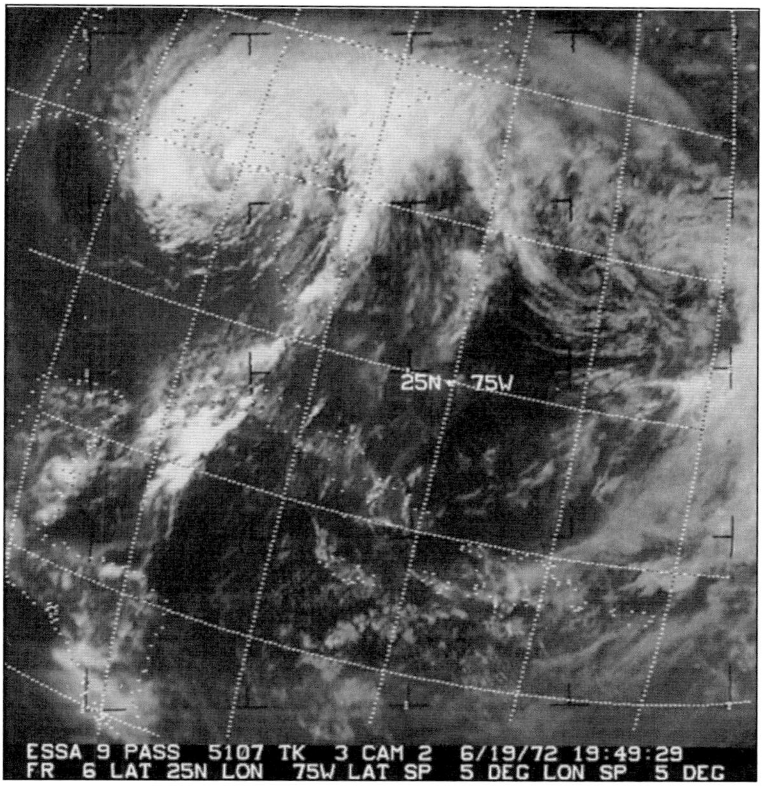

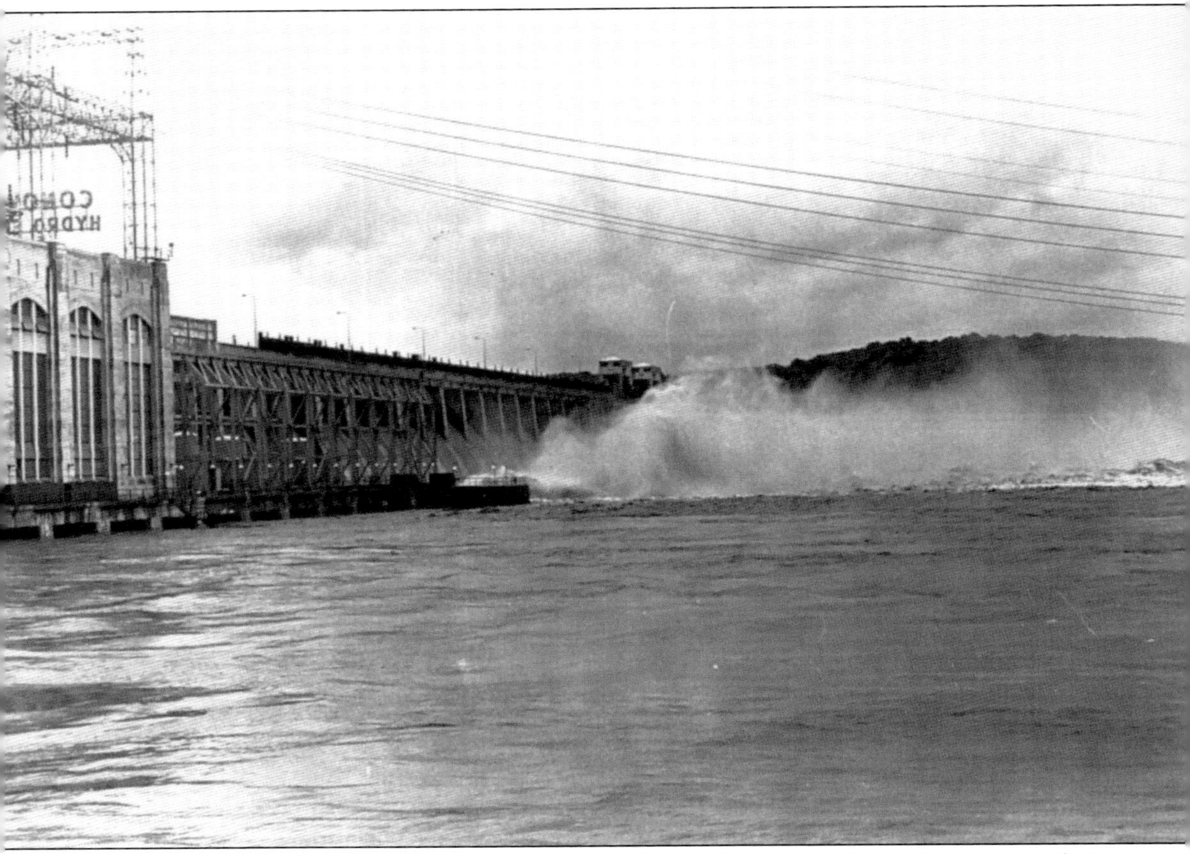

On June 22, the operator at Conowingo, watching the pond gauge rise, knew there was no choice but to calculate how many gates to pull. When employee Dave Goodwin arrived on the headworks that day, he climbed into the cab of a gantry crane and received an order from the control room he had never heard before. "Pull ten gates!" the voice said. Shocked, Goodwin radioed back to verify, "You mean pull gate No. 10?" The reply, "No! Pull gates one through ten!" One by one, those 10 crest gates, and then 40 more, were pulled, as seen in this photograph. Paul Hoopes, the crane operator who pulled the last gate, was asked, "What did you feel like after you opened the 50th gate?" He said, "I wished we had more gates!" Nothing more could be done to control the river as the pond continued to rise above elevation 108.5. (Courtesy of the US Army Corps of Engineers.)

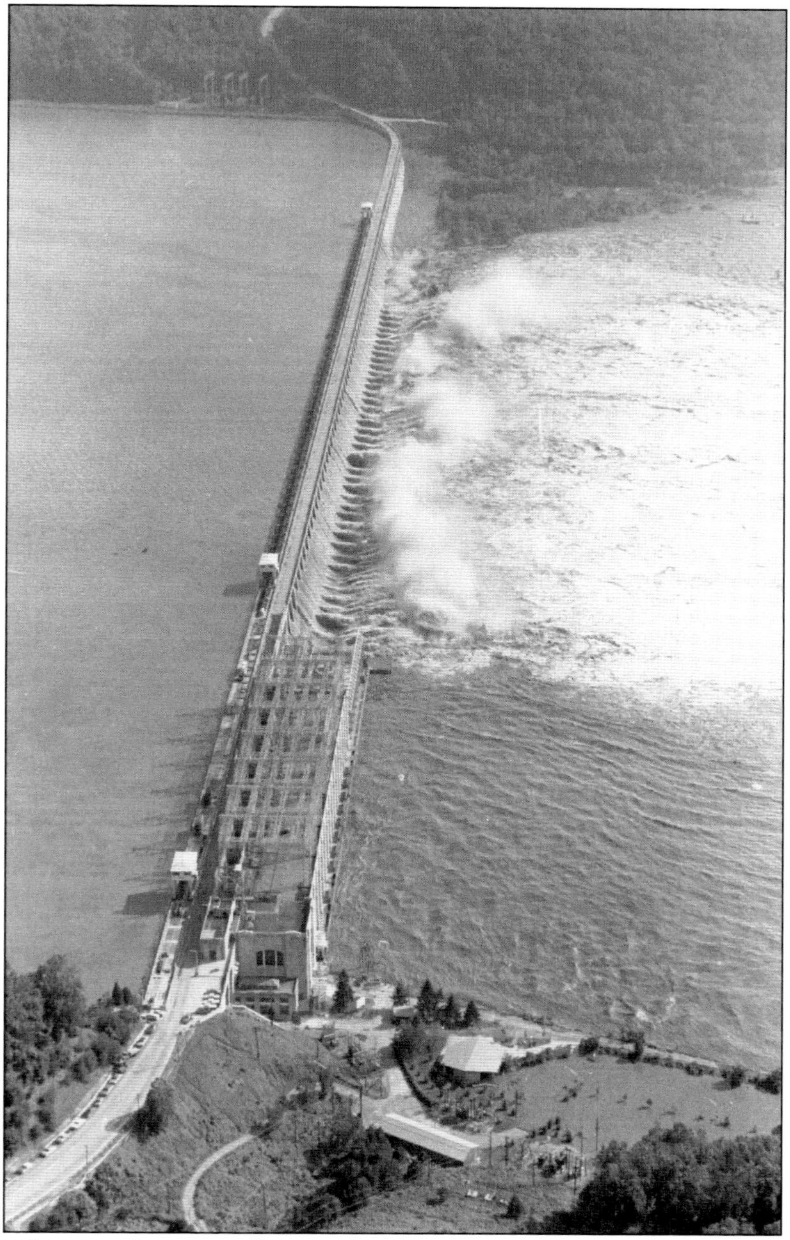

With its floodgates wide open and water continuing to rise, the dam was in danger of being overtopped, but another more alarming possibility also loomed. As a concrete gravity dam, Conowingo sat in place under the force of its own weight. There was a fear that the swollen river might lift the dam from its foundations, destroying a portion of it and wiping out the towns downriver. On June 23 at 10:30 p.m., the superintendent of the dam, Paul K. English, released the following dire warning: "When (the river) reaches 111 feet . . . it will be a hazy area. It is expected to reach this stage within the next 2 or 3 hours. It is not known . . . whether a structure of the dam may give and . . . people downstream should be advised." Larry Walter was on duty during the crisis and recalled, "The force of that water raging through the dam was just unimaginable, and it just kept coming up, coming up, coming up." Nobody knew how high the river would crest or if the dam would hold. (Courtesy of *Lancaster Online*.)

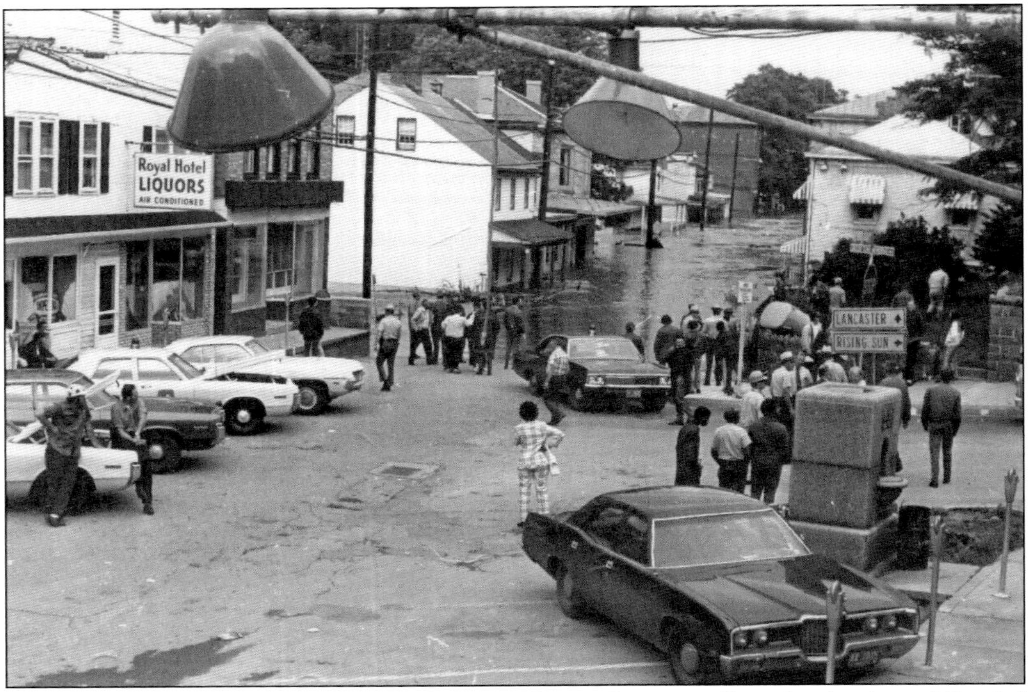

Fed by a week of almost constant rain, the Susquehanna River overflowed its banks and flooded many towns above the Conowingo Dam. People who lived below it also paid a heavy price. In Port Deposit, Maryland, school buses rolled into town and all of the citizens were ordered to leave. More than 3,000 residents were evacuated from Port Deposit, Perryville, and Havre de Grace. These June 1972 images show the floodwaters once again inundating Port Deposit, a town that had endured countless such disasters before. Above is a view of Main Street looking toward the low-lying end of town to the north. The below photograph shows four-foot-high waters flowing down Main Street and skimming the porches of its 19th-century wooden homes. (Both, courtesy of Donald Poist.)

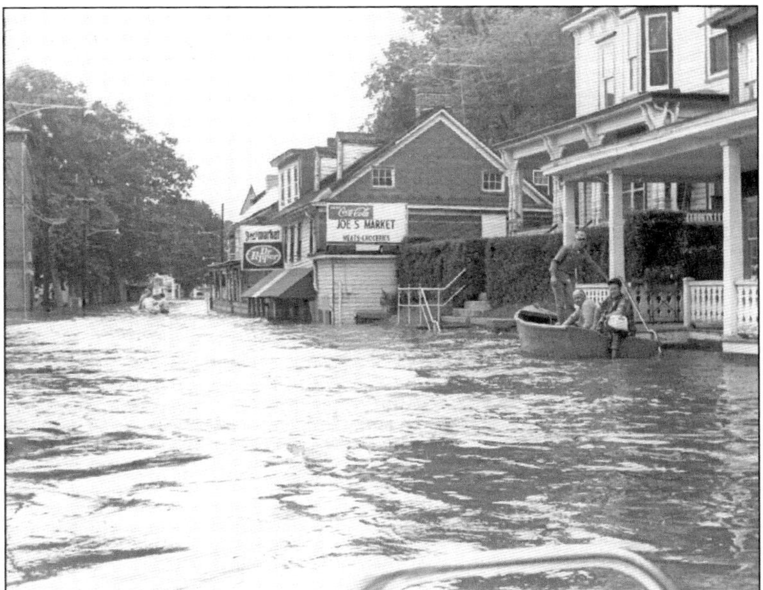

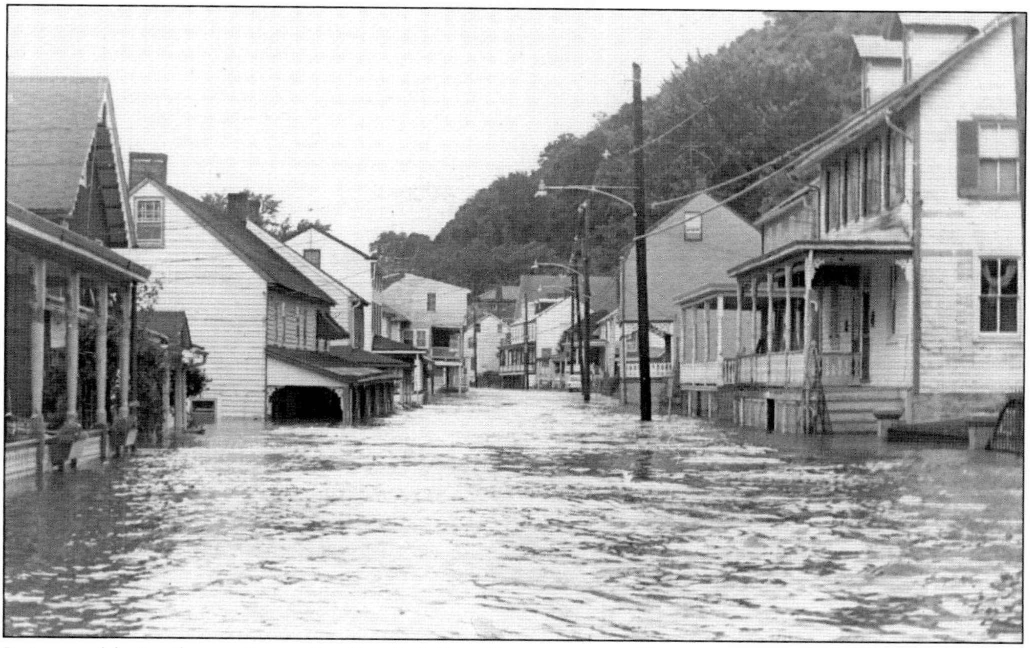

It is notable in these views of a flooded Port Deposit, as well as in photographs of the storm at Conowingo Dam, that the skies are overcast but seldom raining. Hurricane Agnes was unique in that most of the rainfall was far upriver while the storm hovered over Pennsylvania and New York State. Those who worked at the dam and lived in the nearby towns knew the deluge was coming down the watershed; the only thing they could do was brace for its impact and hope for the best. Port Deposit had been flooded so often that most residents knew that to brace for impact meant moving their furniture and valuables to higher floors to keep them dry. This time, however, there were warnings that the very dam structure itself might fail, and if it did, the whole town would be destroyed. (Both, courtesy of Donald Poist.)

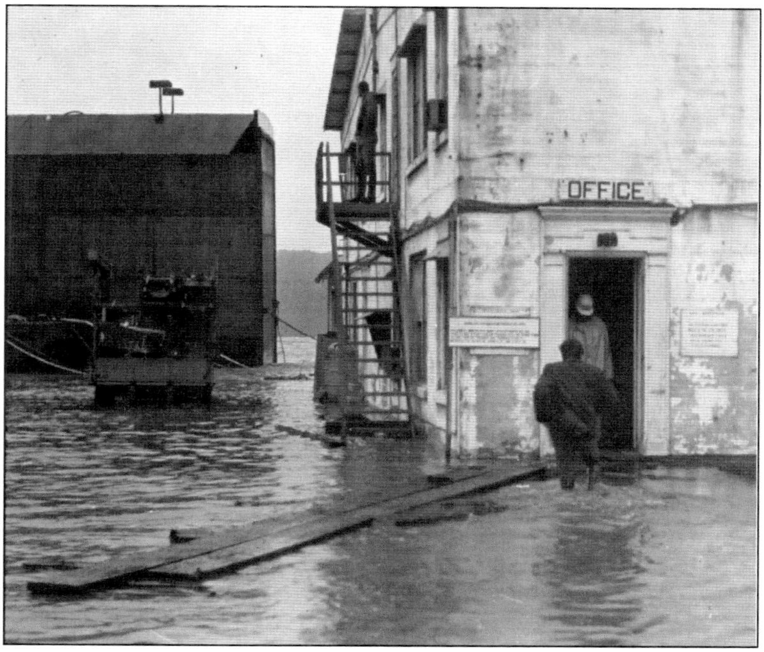

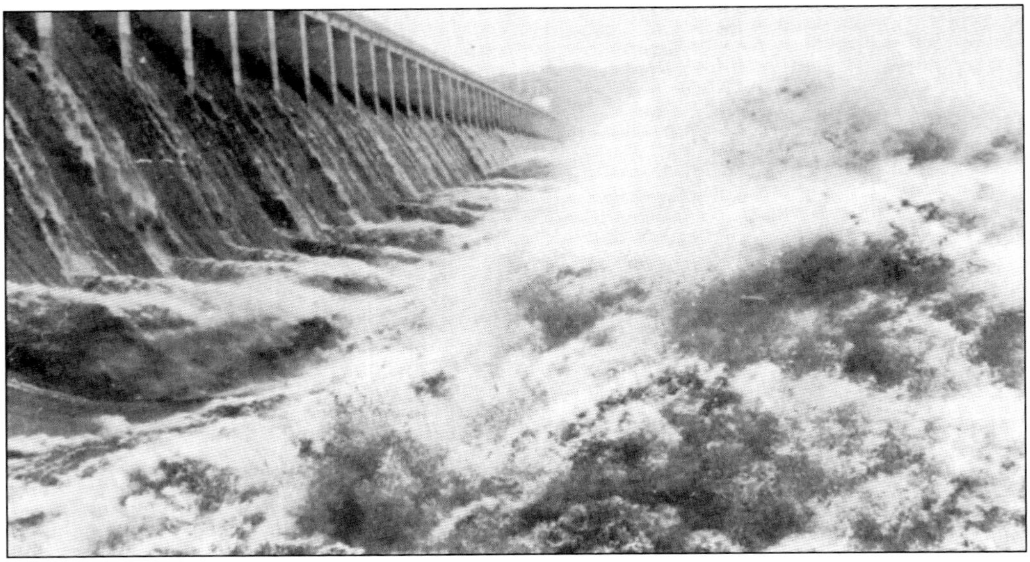

Todd Holden clicked this image while on assignment for the *Aegis* newspaper in Harford County, Maryland. Its caption reads, "Thundering cauldron at the base of the Conowingo Dam. The mud-laden spray of 9 million gallons a second and the thundering roar of water on the rampage afforded this breathtaking glimpse of the Susquehanna River as it has never been seen before." (Courtesy of the *Aegis*.)

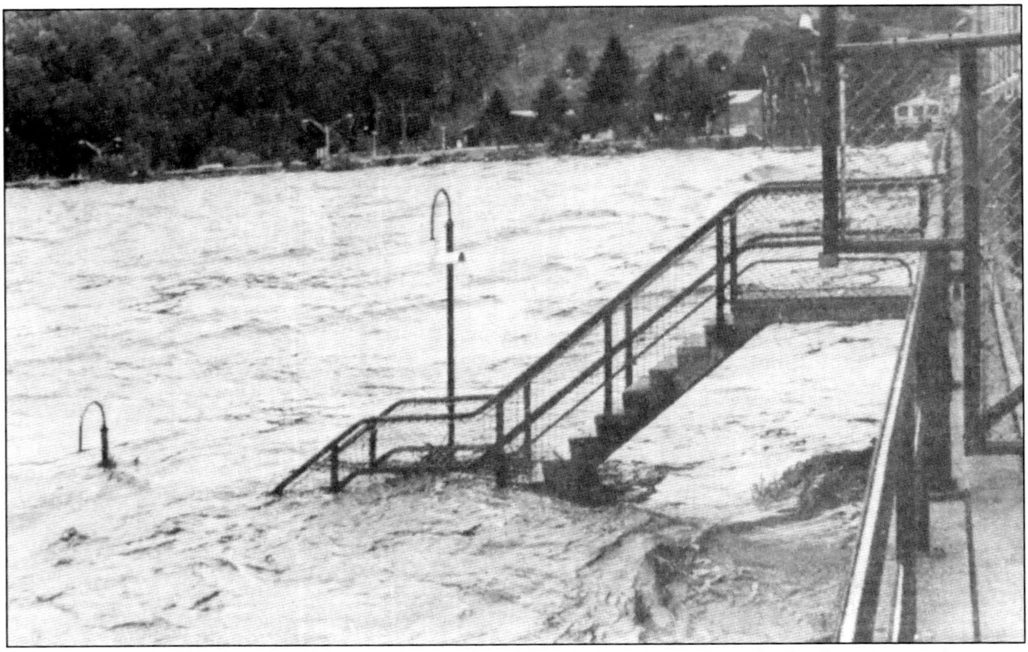

Holden felt the dam trembling as he took this shot, observing, "The whole thing was vibrating, just shaking. And I thought, 'Good God!' And then I saw a part of a cabin go whoosh!, down the spillway." Debris of every description was swept downriver during Hurricane Agnes. Dam workers watched boats and whole trees, 30 and 40 feet long, being thrust through the dam at high velocity and pulverized in the violent churn below. (Courtesy of the *Aegis*.)

Conowingo was built to withstand a stream flow of 800,000 cubic feet per second. These handwritten field notes were taken at 1:30 in the morning on June 24, 1972, by the staff of the US Geological Survey. The notes show a maximum streamflow discharge measurement of 1,128,000 cubic feet per second. At that early morning hour, Conowingo was holding against a far greater river flow than it had been designed for. (Courtesy of the US Geological Survey.)

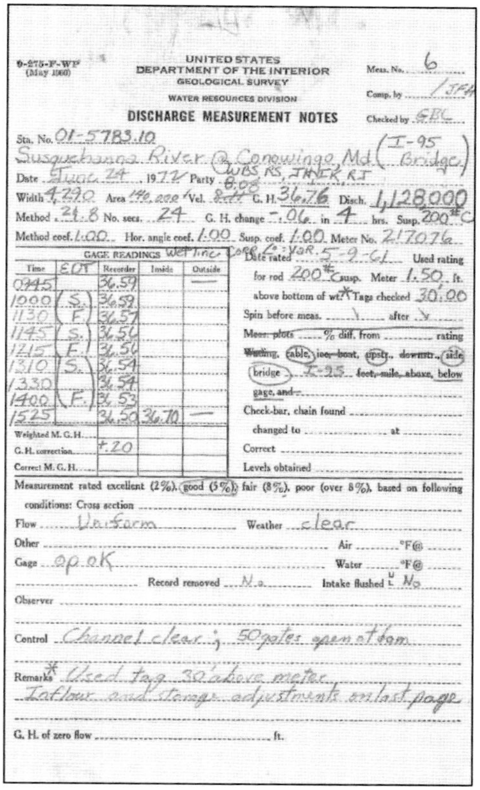
9-275-F-WP (May 1960)

UNITED STATES
DEPARTMENT OF THE INTERIOR
GEOLOGICAL SURVEY
WATER RESOURCES DIVISION

DISCHARGE MEASUREMENT NOTES

Meas. No. 6
Comp. by
Checked by

Sta. No. 01-5783.10
Susquehanna River @ Conowingo, Md. (I-95 Bridge)
Date June 24 1972 Party
Width 4290 Area 140,000 Vel. G. H. 36.76 Disch. 1,128,000
Method No. secs. 24 G. H. change -.06 in 4 hrs. Susp. 200# C
Method coef. 1.00 Hor. angle coef. 1.00 Susp. coef. 1.00 Meter No. 217076

Time		Recorder	Inside	Outside
0945		36.59		—
1000	(S.)	36.59		
1130	(F.)	36.57		
1145	(S.)	36.56		
1215	(F.)	36.56		
1310	(S.)	36.54		
1330		36.54		
1400	(F.)	36.53		
1525		36.50	36.70	—
Weighted M. G. H.				
G. H. correction		+.20		
Correct M. G. H.				

Date rated 5-9-61 Used rating for rod 200# susp. Meter 1.50 ft. above bottom of wt. Tags checked 30.00
Spin before meas. after
Meas. plots % diff. from rating
Wading, cable, ice, boat, upstr., downstr., side bridge I-95 feet, mile, above, below gage, and
Check-bar, chain found
changed to at
Correct
Levels obtained

Measurement rated excellent (2%), good (5%), fair (8%), poor (over 8%), based on following conditions: Cross section
Flow Uniform Weather clear
Other Air °F@
Gage op. OK Water °F@
Record removed No Intake flushed No
Observer
Control Channel clear; 50 gates open at dam
Remarks * Used tag 30' above meter
Inflow and storage adjustments on last page
G. H. of zero flow ft.

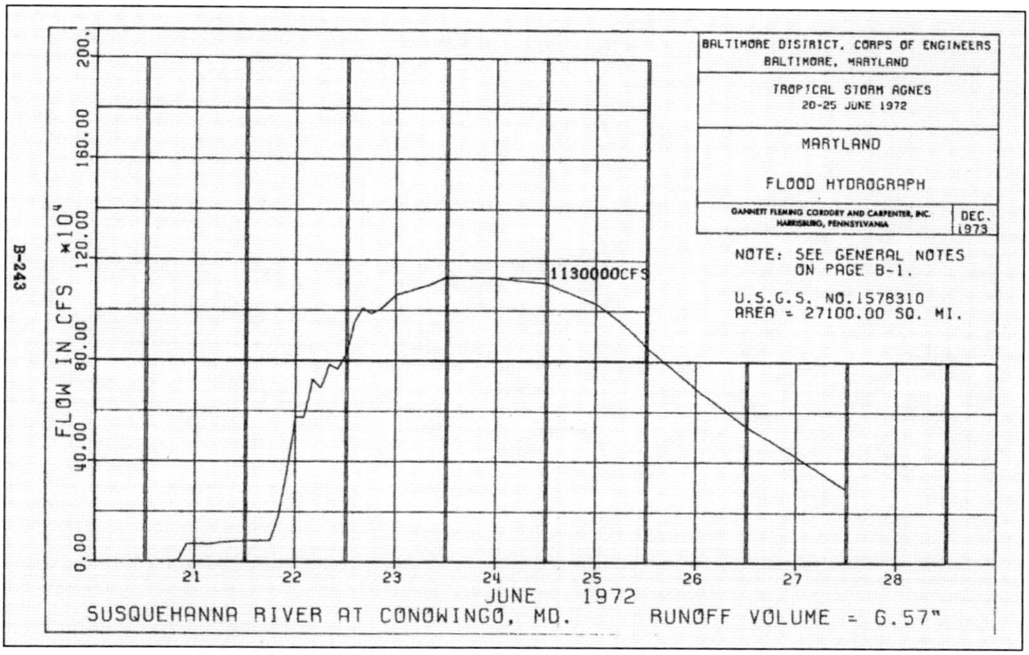

This hydrograph shows the precipitous rise of the river during a roughly 48-hour period, when streamflow soared from 20,000 cubic feet per second to over one million cubic feet per second. With all 50 crest gates open, Conowingo was spewing out more than 7.5 million gallons of muddy water a second. (Courtesy of the US Geological Survey.)

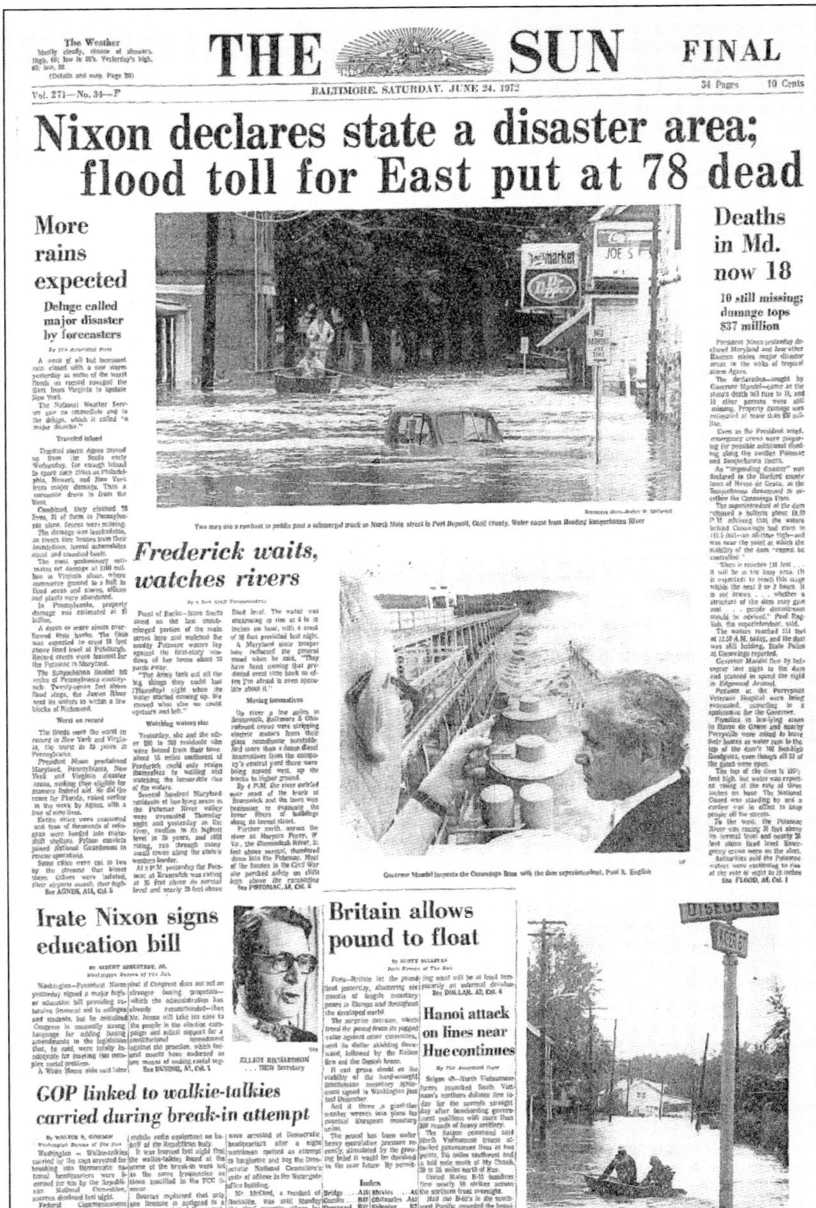

THE SUN FINAL

Vol. 271—No. 34—F BALTIMORE, SATURDAY, JUNE 24, 1972 34 Pages 10 Cents

Nixon declares state a disaster area; flood toll for East put at 78 dead

More rains expected

Deluge called major disaster by forecasters

Deaths in Md. now 18

10 still missing; damage tops $37 million

Frederick waits, watches rivers

Irate Nixon signs education bill

GOP linked to walkie-talkies carried during break-in attempt

Britain allows pound to float

Hanoi attack on lines near Hue continues

President Nixon toured the flooded region and declared it a disaster area. Maryland governor Marvin Mandel flew to the dam by helicopter to assess the imminent danger with Conowingo superintendent Paul K. English. Unsubstantiated speculation has circulated for years that the state government had a contingency plan to blow up part of the dam to let more floodwaters pass through rather than allowing the river to breach the dam or, more catastrophically, to topple it. A worker who was at the dam testified that he drilled holes for the explosives, but that the charges were never set. With water behind the dam cresting near the top, to elevation 111.5, a dam engineer said, "We were working with safety factors at levels where we had never been before." Maj. General Edwin Warfield of the Maryland National Guard said, "20 helicopters were positioned in the area to evacuate residents below the dam if it should crack." (Courtesy of the *Baltimore Sun*.)

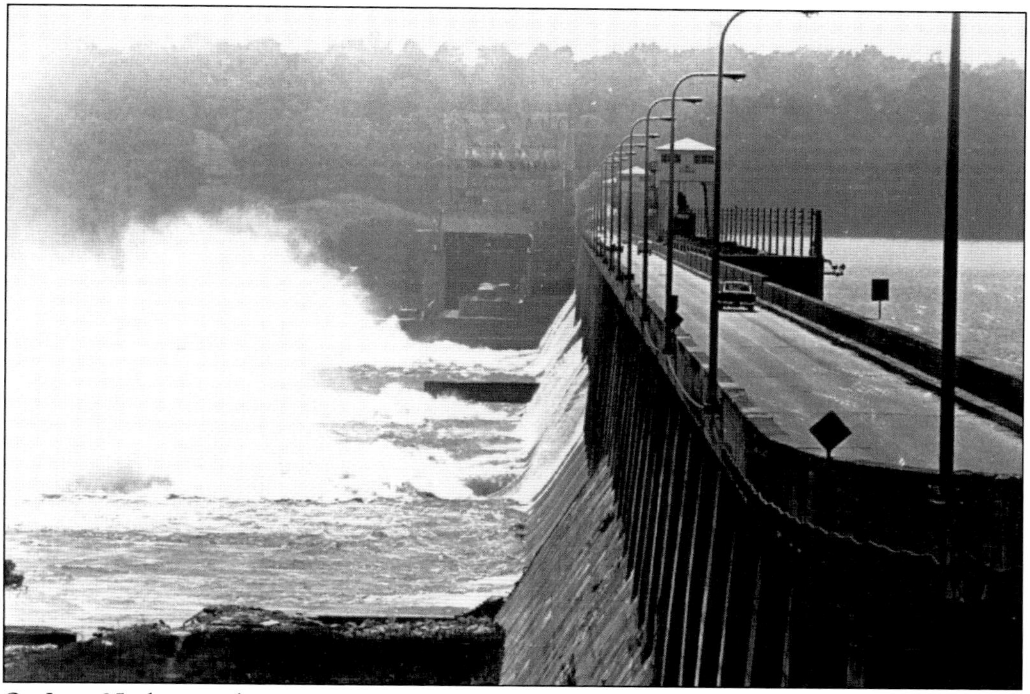

On June 25, the river began to subside and dam workers breathed a sigh of relief. The 44-year-old Conowingo Dam had survived the storm, and the fears of the downriver public were alleviated. As the waters receded, the residents of flood-ravaged towns returned by boat or on foot to reclaim what was left of their homes and valuables. It was time to assess the damage and undertake the messy chore of pushing the mud back into the river. As the doleful narrator of a government film that was made on the hurricane laments, "Everywhere mud. Sour, stinking mud." The total damage caused by Hurricane Agnes was estimated at well over $1 billion and the storm was at that time the costliest hurricane ever to hit the United States. (Above, courtesy of the US Army Corps of Engineers; right, courtesy of the *Baltimore Sun*.)

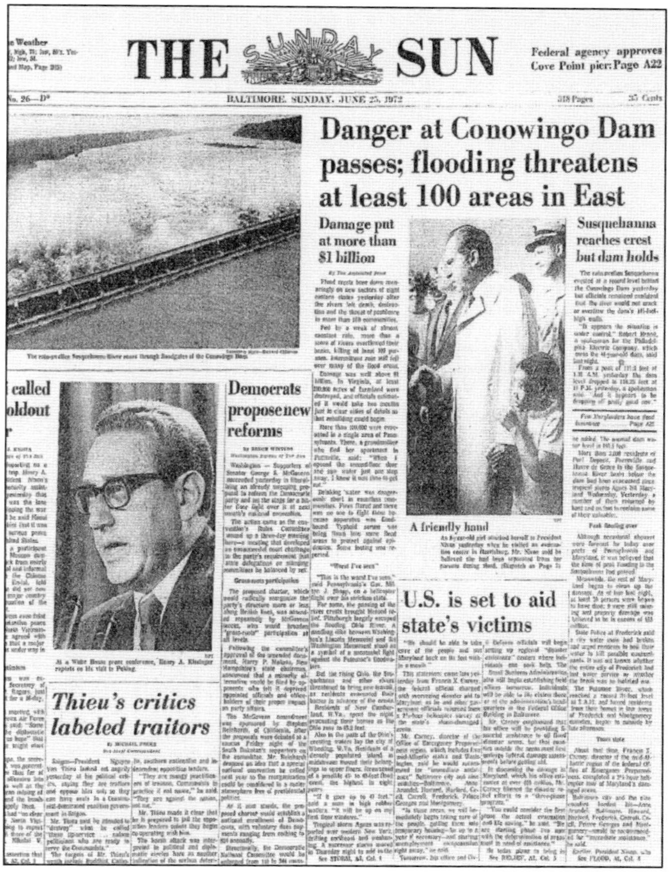

THE SUNDAY SUN

Federal agency approves Cove Point pier: Page A22

No. 26—D* BALTIMORE, SUNDAY, JUNE 25, 1972 318 Pages 25 Cents

Danger at Conowingo Dam passes; flooding threatens at least 100 areas in East

Damage put at more than $1 billion

Susquehanna reaches crest but dam holds

Democrats propose new reforms

A friendly hand

U.S. is set to aid state's victims

Thieu's critics labeled traitors

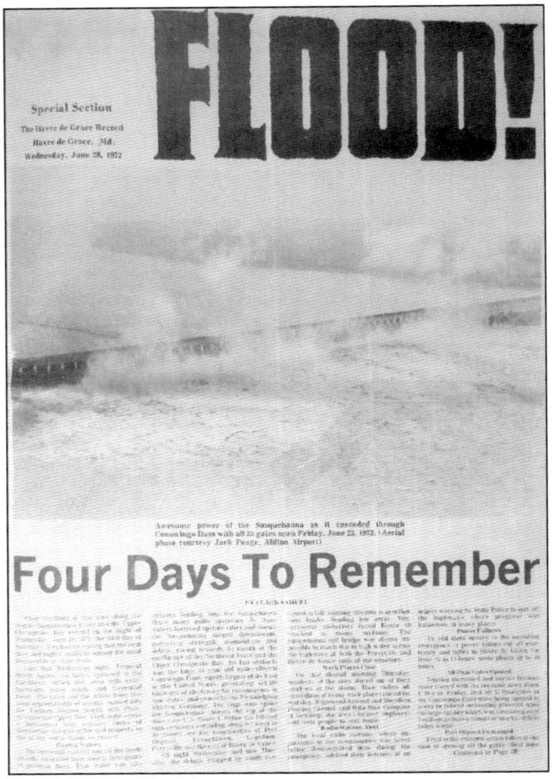

FLOOD!

Special Section
The Havre de Grace Record
Havre de Grace, Md.
Wednesday, June 28, 1972

Awesome power of the Susquehanna as it cascaded through Conowingo Dam with all 53 gates open Friday, June 23, 1972. (Aerial photo courtesy Jack Paage, Aldino Airport)

Four Days To Remember

Conowingo Dam had its own post-Agnes mess to clean up. During the worst of the flood, the dam operators ran all 11 turbine generators in an attempt to move every gallon they could through the powerhouse and lessen the force against the dam. Power was eventually shut off during the storm, but not before the penstocks and scroll cases had become jammed with tree trunks and debris. After Agnes, it took maintenance workers six months to unwater the draft tubes and cut up the debris with chain saws. Then in 1978, in order to eliminate the possibility of the dam shifting on its foundation, the dam structure was actually anchored into the bedrock. A total of 537 holes were drilled through the structure and into the foundation, and stranded wire tendons were then installed to pin the dam down and increase its stability against future storms. (Left, courtesy of *Havre de Grace Record.*)

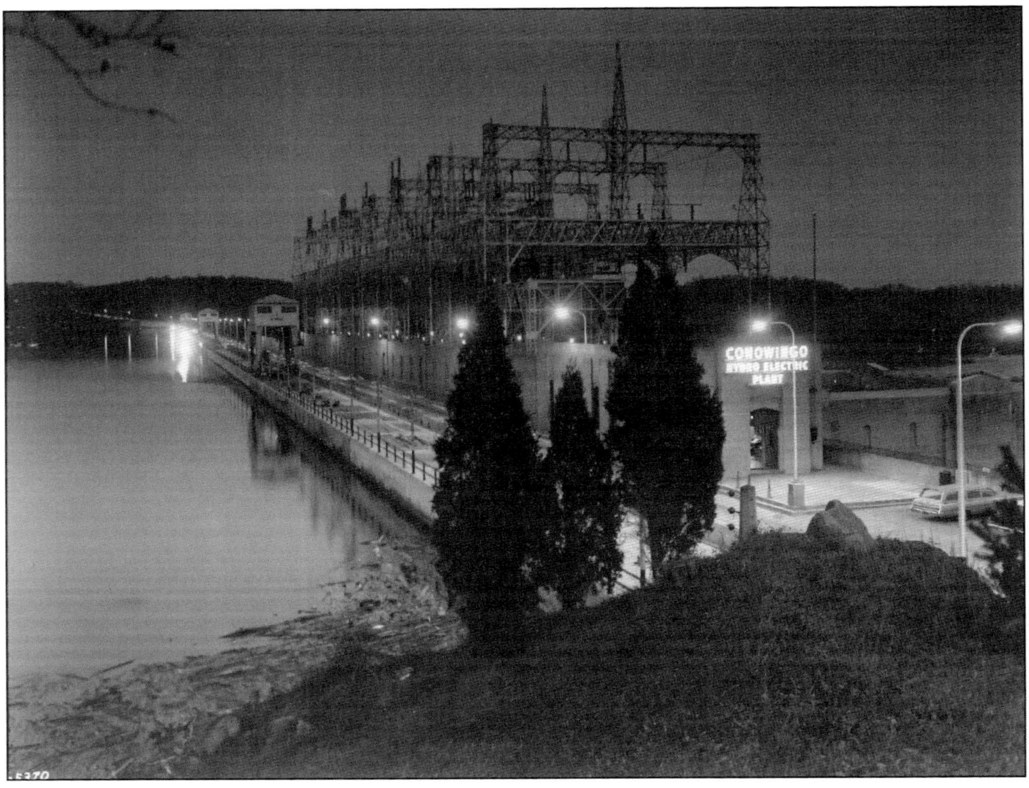

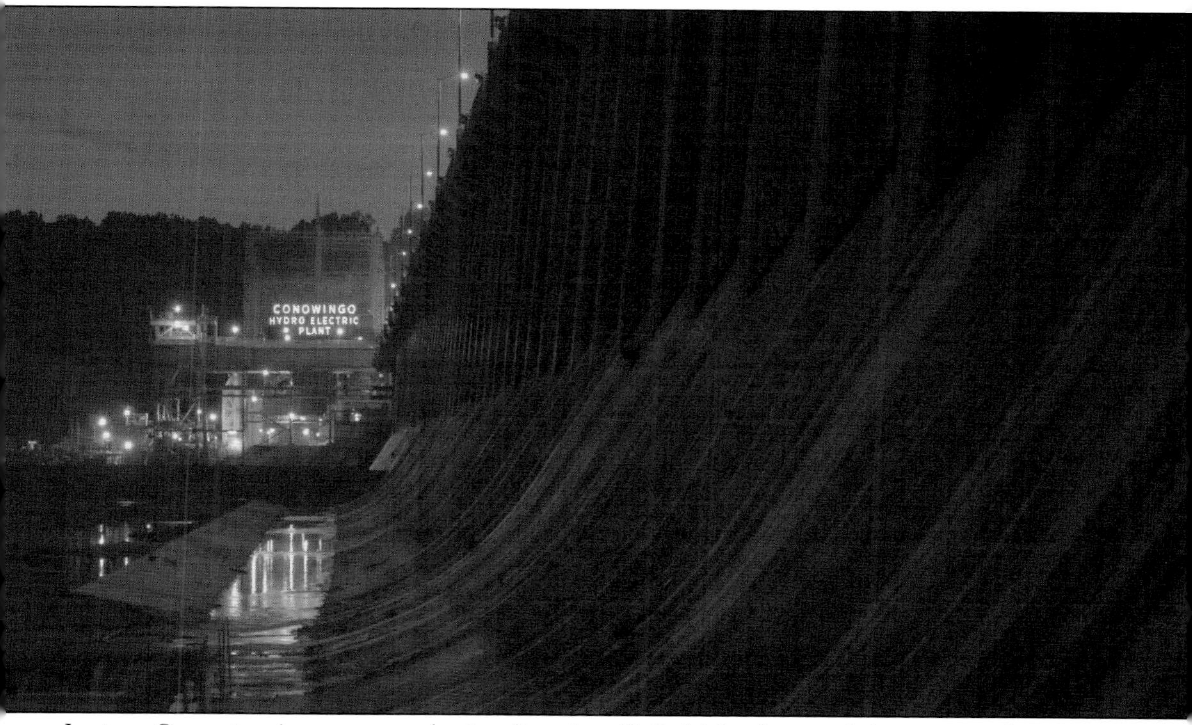

In time, Conowingo's power ceased going directly to Philadelphia but was fed to a regional power interconnection system, colloquially known as "the grid," that served the entire eastern United States. As Conowingo aged, it was eventually dwarfed by the power-generating capacities of coal, natural gas, and nuclear power plants. However, Conowingo remained vital to the safety and security of the East Coast power grid in that, in the event of a widespread system failure, the dam could jump-start the electric distribution system, a capacity known as "black start." Because water turned its turbines, Conowingo could always open a gate, bring its generators online, and send out power. Power plants that did not have the capability to start themselves—coal, nuclear, or natural gas—could then reignite using the electricity originating at Conowingo. During major East Coast blackouts, it was Conowingo Dam that helped reestablish the power grid, an honor that inspired staff engineer Tom Jenkins to christen it "the little dam that could."